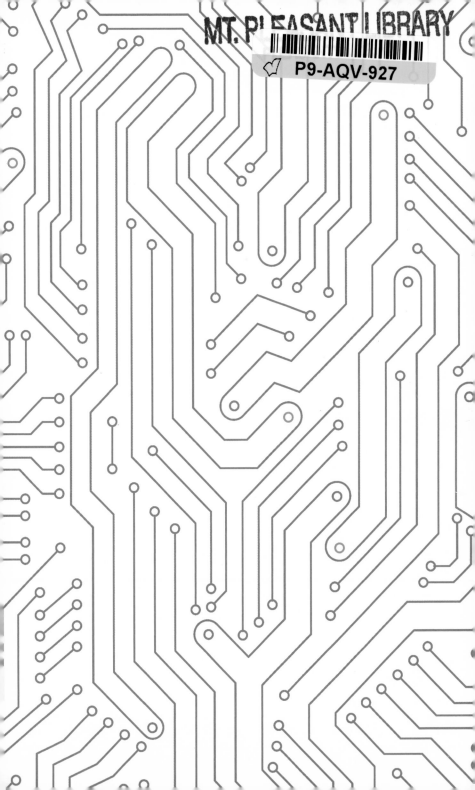

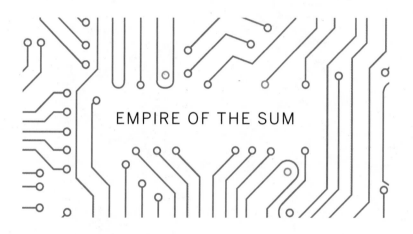

EMPIRE OF THE SUM

ALSO BY KEITH HOUSTON

Shady Characters:
The Secret Life of Punctuation, Symbols &
Other Typographical Marks

The Book:
A Cover-to-Cover Exploration of
the Most Powerful Object of Our Time

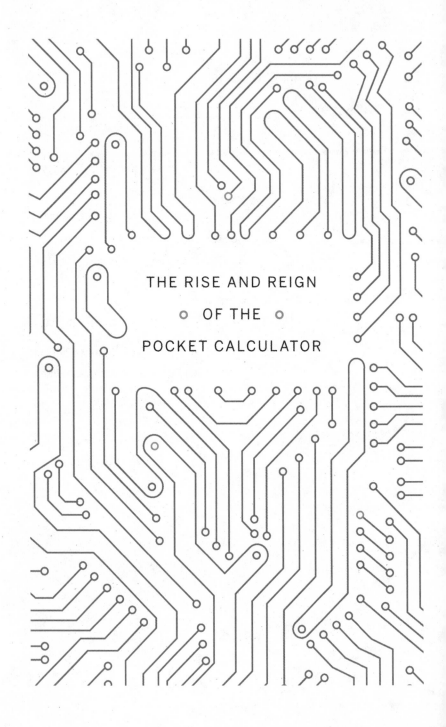

THE RISE AND REIGN

o OF THE o

POCKET CALCULATOR

Empire

OF THE Sum

KEITH HOUSTON

W. W. NORTON & COMPANY

Celebrating a Century of Independent Publishing

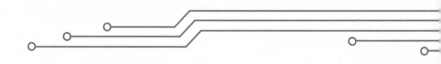

For information about permission to reproduce selections from this book, write to
Pemissions, W. W. Norton & Company, Inc., 500 Fifth Avenue, New York, NY 10110

For information about special discounts for bulk purchases, please contact
W. W. Norton Special Sales at specialsales@wwnorton.com or 800-233-4830

Manufacturing by Versa Press
Book design by Abbate Design
Production manager: Anna Oler

ISBN: 978-0-393-88214-8

W. W. Norton & Company, Inc.
500 Fifth Avenue, New York, N.Y. 10110
www.wwnorton.com

W. W. Norton & Company Ltd.
15 Carlisle Street, London W1D 3BS

1 2 3 4 5 6 7 8 9 0

CONTENTS

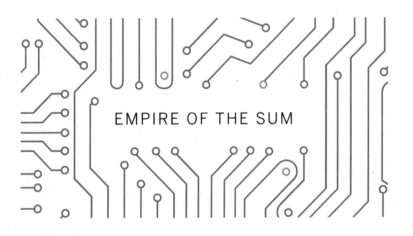

EMPIRE OF THE SUM

INTRODUCTION

THE POCKET CALCULATOR is predicated on two things: calculation and pockets. And though humanity can rightfully claim to having placed the one into the other, when it comes to inventing them, nature got there first. Both times.

Consider, briefly, the pocket. The oldest known pocket worn by a human, in the form of a leather pouch strung onto a belt, is at least five thousand years old. This prehistoric receptacle was discovered in 1991 when a pair of hikers walking in the Italian Ötztal Alps came across a slumped, emaciated corpse entombed in glacial ice. At first thought to be a music teacher who had gone missing during World War II, the body was later determined to be that of a Neolithic hunter who had been wounded by a blow to the head and left on the mountain to die many thousands of years earlier. At the time of his death, Ötzi the Iceman, as he was christened, wore a leather belt and a pouch containing flint and bone tools and a piece of a fungus often used as tinder.[1] Some latter-day commentators have maligned Ötzi's pouch as a fanny pack, but let us not be distracted by such name-calling: this was a pocket, and an ancient one at that.[2]

Yet the clever use of gravity to hold things that might otherwise escape is not a human invention. Pockets have spontaneously manifested across a variety of species, phyla, and even kingdoms, with kangaroos, pitcher plants, and sea otters being only a few of nature's more charismatic possessors of pockets. As a case in point, prehistoric marsupials called *Peratherium* are known to have lived in North America around 45 million years ago, making animal pockets at least nine thousand times older than humanity's first known effort.[3] Nature laughs at Ötzi's pocket.

Perhaps more surprising than *Homo sapiens'* belated discovery of pockets is that many animal species can count and, almost certainly, have been counting since long before humans existed. Scientists have conducted experiments on baboons, for example, in which peanuts were inserted into two separate caches so that the monkeys could not see the overall amounts of food in each cache. The only way the baboons could have selected the better-stocked larder—which they did, almost three-quarters of the time—was to have counted the food parcels as they were inserted.[4]

Nor is counting limited to mammals. Ravens and other corvids are famously intelligent, having been seen in the wild to use tools (and perhaps even weapons) and having also solved logical puzzles in experimental contexts.[5] In the first half of the twentieth century, German zoologist Otto Köhler took things a stage further by demonstrating that ravens could count up to seven. Pigeons managed five, he recorded, budgerigars and jackdaws six.[6] Later, in the 1980s and 1990s, an animal psychologist named Irene Pepperberg taught Alex, an African Grey parrot, to count up to six. Not much for ravens to worry about, then, except that Alex also mastered the concept of "none," or zero, an idea that humanity had not seriously investigated until the third or second century BCE.[7]

Fish count too. Experiments with guppies demonstrated that they preferred to join larger groups of the same species. Intriguingly, the guppies also showed hints of what is called *numerosity*—not the ability to count, exactly, but rather to estimate relative sizes or quantities where counting individual items is no longer practical.[8] When presented with small groups of up to four fish each, the guppies were able to consistently distinguish between them, but beyond that, bigger groups had to be at least twice as large as smaller groups for the guppies to tell the difference.[9]

Turning to even smaller creatures, some ants can communicate numeric quantities to one another, exploring mazes and instructing their comrades to follow the same paths by counting off junctions.[10] Some spiders also count, although *Portia africana*, a central African jumping spider that has been shown to hesitate when presented with an unexpected number of prey, seems to lose track after the number two.[11]

There is something primal about counting, in other words, independent as it is of possessing a language, a large brain, or even a backbone. And if we can observe it in today's animals, it is likely to have arisen in their ancestors millions of years in the past as an evolutionary response to the natural world.[12] That same evolutionary response has shaped humans too: babies look longer at sets of objects from which some have been unexpectedly removed, suggesting that they must have counted those objects and they are surprised when some are missing.[13] If (and forgive the reach) the pocket is a natural consequence of gravity, counting might be said to be a natural consequence of being alive. To live is to count, and to count is to calculate.

Of course, human beings are nothing if not contrary. Endowed with a preternatural facility for calculation in both our conscious and subconscious minds, we have nevertheless spent most of recorded history trying to raise ourselves above this most fundamental part of the human condition. Why count, we asked ourselves for millennium after millennium, when a machine could do it for us? Increment by increment, we arrived at an answer. There exist pockets; there exists counting. For a brief, halcyon moment at the tail end of the twentieth century, we put one into the other, and everyone took notice.

1 THE HAND

IGURING OUT WHEN humans began to count systematically, with purpose, is not easy. Our first real clues are a handful of curious, carved bones dating from the final few millennia of the three-million-year expanse of the Old Stone Age, or Paleolithic era. Those bones are humanity's first pocket calculators: For the prehistoric humans who carved them, they were mathematical notebooks and counting aids rolled into one. For the anthropologists who unearthed them thousands of years later, they were proof that our ability to count had manifested itself no later than 40,000 years ago.[1]

In 1973, while excavating a cave in the Lebombo Mountains, near South Africa's border with Swaziland, Peter Beaumont found a small, broken bone with twenty-nine notches carved across it. The so-called Border Cave had been known to archaeologists since 1934, but the discovery during World War II of skeletal remains dating to the Middle Stone Age heralded a site of rare importance. It was not until Beaumont's dig in the 1970s, however, that the cave gave up its most significant treasure: the earliest known *tally stick*, in the form of a notched, three-inch long baboon fibula.[2]

On the face of it, the numerical instrument known as the tally stick is exceedingly mundane. Used since before recorded history—still used, in fact, by some cultures—to mark the passing days, or to account for goods or monies given or received, most tally sticks are no more than wooden rods incised with notches along their length. They help their users to count, to remember, and to transfer ownership.[3] All of which is reminiscent of writing, except that writing did not arrive until a scant 5,000 years ago—and so, when the Lebombo bone was determined to be some 42,000 years old, it instantly became one of the most intriguing archaeological artifacts ever found.[4] Not only does it put a date on when *Homo sapiens* started counting, it also marks the point at which we began to delegate our memories to external devices, thereby unburdening our minds so that they might be used for something else instead. Writing in 1776, the German historian Justus Möser knew nothing of the Lebombo bone, but his musings on tally sticks in general are strikingly apposite:

> The notched tally stick itself testifies to the intelligence of our ancestors. No invention is simpler and yet more significant than this.[5]

It is not clear what quantity the twenty-nine notches carved into the Border Cave's baboon fibula represents. It *is* a number, that much is known: had the bone been purely decorative, the notches would have been added all at once, but four different tools were used over time to add to the count.[6] As such, the Lebombo bone is likely to be the earliest mathematical device ever found. (Sadly, it is too great a leap to call it the earliest known pocket calculator. Humans started wearing clothes around 170,000 years ago, but pockets themselves, as evidenced by Ötzi's pouch, are probably no more than a few thousand years old.)[7]

If the Lebombo bone answers the question, at least partly, of

O The Lebombo bone, a baboon fibula bearing twenty-nine notches carved by four separate tools and dating to the Middle Stone Age, is thought to be the oldest known mathematical artifact. *McGregor Museum, Kimberley, South Africa. Photo: Robert Hart.*

when humans learned to count, it leaves another one unanswered: How did they learn to do so?

COUNTING, FUNDAMENTALLY, is the act of assigning distinct labels to each member of a group of similar things to convey either the size of that group or the position of individual items within it.[8] The first type of counting yields cardinal numbers such as "one," "two," and "three"; the second gives ordinals such as "first," "second," and "third."

At first, our hominid ancestors probably did not count very high. Many body parts present themselves in pairs—arms, hands, eyes, ears, and so on—thereby leading to an innate familiarity with the concept of a pair and, by extension, the numbers 1 and 2. But when those hominids regarded the wider world, they did not yet find a need to count much higher. One wolf is manageable; two wolves are a challenge; any more than that and time spent counting wolves is better spent making oneself scarce. The result is that the very smallest whole numbers have a special place in human culture, and especially in language. English, for instance, has a host of specialized terms centered around *twoness*: a brace of pheasants; a team of horses; a yoke of oxen; a pair of, well, anything. An ancient Greek could employ specific plurals to distinguish between groups of one, two, and many friends (*ho philos, to philo,* and *hoi philoi*).[9] In Latin, the numbers 1 to 4 get special treatment, much as "one" and "two" correspond to "first" and "second," while "three" and "four" correspond directly with "third" and "fourth." The Romans extended that special treatment into their day-to-day lives: after their first four sons, a Roman family would typically name the rest by number (Quintus, Sextus, Septimus, and so forth), and only the first four months of the early Roman calendar had proper names. Even tally marks, the age-old "five-barred gate" (卌) used to score card games or track rounds of drinks, speaks of a deep-seated need to keep things simple.[10]

Counting in the prehistoric world would have been intimately bound to the actual, not the abstract. Some languages still bear traces of this: a speaker of Fijian may say *doko* to mean "one hundred mulberry bushes," but also *koro* to mean "one hundred coconuts." Germans will talk about a *Faden*, meaning a length of thread about the same width as an adult's outstretched arms.[11] The Japanese count different kinds of things in different ways: there are separate sequences of cardinal numbers for books; for other bundles of paper such as magazines and newspapers; for cars, appliances, bicycles, and similar machines; for animals and demons; for long, thin objects such as pencils or rivers; for small, round objects; for people; and more.[12]

Gradually, as our day-to-day lives took on more structure and sophistication, so, too, did our ability to count. When farming a herd of livestock, for example, keeping track of the number of one's sheep or goats was of paramount importance, and as humans divided themselves more rigidly into groups of friends and foes, those who could count allies and enemies had an advantage over those who could not.[13] Number words graduated from being labels for physical objects into abstract concepts that floated around in the mental ether until they were assigned to actual things.

Even so, we still have no real idea how early humans started to count in the first place. Did they gesture? Speak? Gather pebbles in the correct amount? To form an educated guess, anthropologists have turned to those tribes and peoples isolated from the greater body of humanity, whether by accident of geography or deliberate seclusion. The conclusion they reached is simple. We learned to count with our fingers.[14]

A 1913 SURVEY OF the number words used by several Native American tribes found that many of those words were related to "finger," "thumb," and "hand." Counterintuitively, perhaps, despite the general possession of ten fingers per person, fewer than half of those tribes counted in multiples of ten. About a third used systems that revolved around the number 5, which was often referred to as "fingers finished," "all finished," "gone," or "spent." A further tenth of the tribes used *vigesimal* schemes based on the number 20 ("all hands and feet"), while a few contrarian outliers used 2-, 3-, and 4-based systems with less obvious connections to human anatomy.[15]

Fifteen years earlier, a group of scientists from Cambridge, England, had made a series of visits to the islands of the Torres Straits, strung between Papua New Guinea to the north and Australia to the south.[16] A. C. Haddon, the driving force behind the expeditions, recounted

> There was another system of counting by commencing at the little finger of the left hand, *kotodimura*, then following on with the fourth finger, *kotodimura gorngozinga* (or *quruzinger*); middle finger, *il get*; index finger, *klak-nĕtoi-gĕt*; thumb, *kabaget*; wrist, *perta* or *tiap*; elbow joint, *kudu*; shoulder, *zugukwoik*; left nipple, *susu madu*; sternum, *kosa, dadir*; right nipple, *susu madu*, and ending with the little finger of the right hand.[17]

In this way, Haddon said, starting on one side of the body and traversing over to the other, the islanders could count to nineteen. More recently, a math teacher named Glen Lean catalogued the number words for 883 of the 1,200 known languages from Papua New Guinea and Micronesia and found that the use of fingers for counting was foundational to many of those languages. Like the Torres Strait islanders, the Papua New Guineans then carried on to the forearm, elbow, eyes, nose, ears, and other body parts.[18] A study of Yupno, a language indigenous to Papua New Guinea's Finisterre Mountain range, recorded that Yupno men added their testicles and penis for good measure, allowing them to count to thirty-three using body parts alone.[19]

Hold my earliest attested beer, an ancient Sumerian might have said.[20]

From the sixth millennium onward, the valley between the Tigris and the Euphrates Rivers—Mesopotamia, the ancient Greeks called it, the "land between rivers"—harbored one of the world's earliest civilizations.[21] Having mastered animal husbandry and the cultivation of crops, Mesopotamian farmers became the engine of a new agrarian economy. Almost from the beginning, it seems, they used small clay tokens an inch or so in size, hand-rolled into the shape of spheres, cones, disks, and other simple shapes, to keep records.[22] Each shape stood for a fixed quantity of some good or other: A cone represented a small quantity of cereal, a sphere a larger amount, and a flat disk the largest. Ovoids were jars of oil; cylinders and rounded disks were farm animals; and so on.[23]

Around 3300 BCE, as Mesopotamia's scattered farming communities began to coalesce into the patchwork of city-states called Sumer, their use of tokens became more sophisticated. At first, batches of tokens were wrapped in clay balls called *bullae* and marked with personal seals to create records of important transactions. Later, the surfaces of those *bullae* were impressed with the tokens to be sealed inside so that a *bulla*'s contents could be divined without having to break it open. Once it became apparent that the signs on the outside were as useful as the tokens on the inside, the tokens themselves became surplus to requirements—and the signs, says a theory first proposed by French-American archaeologist Denise Schmandt-Besserat, evolved into the distinctive angular form of cuneiform writing.[24]

Cuneiform tablets show that the Sumerians and their successors, the Akkadians and Babylonians, used *sexagesimal* numbers. That is, their numerical system was rooted in the number 60. Whereas decimal gives rise to round numbers such as 1, 10, and 100 (or 10 squared), the Sumerians counted in terms of 1, 60, 3,600 (or 60 squared), and so on. There are practical advantages to this, since 60 can be divided into whole numbers by 1, 2, 3, 4, 5, 6, 10, 12, 15, 20, 30, and 60, but, as E. F. Robertson, late of St. Andrews University in Scotland, points out, it is rare for a culture to *choose* the base for its number system.[25] More often, as illustrated by those Native American tribes, it naturally settles upon a base when it begins to count. Counting on five fingers leads to the quinary system, or base 5; two hands lead to decimal, or base 10; two hands and two feet to base 20, or vigesimal. How, then, did the Sumerians land on base 60?

The answer lies in the Sumerians' tokens and *bullae*. Successive scholars have noted that the shapes made when tokens were pushed into the soft clay of a *bulla* appear to be very similar to the number symbols used on the earliest "proto-literate" clay tablets. That is, the shapes *and* the values of physical tokens seem to have carried over directly to the written sexagesimal numerals used by the earliest literate Sumerians. As such, the ancient Mesopotamians must have been

O   A *bulla*, opened and since repaired, and the tokens that it once enclosed. Made between 3700 and 3200 BCE, the tokens are thought to represent wages for four days' work, four quantities of metal, a single large quantity of barley, and two smaller amounts of some other, unknown commodity. This is one of fewer than 250 extant *bullae*. *The Schøyen Collection MS 4631. The Schøyen Collection, Oslo and London.*

counting in base 60 *on their fingers* long before they, or, indeed, any-one else on the planet, could set out numbers in writing.[26]

The Mesopotamians' unique counting method is thought to come from a mix of a *duodecimal* system that used the twelve finger joints of one hand and a quinary system that used the five fingers of the other. By pointing at one of the left hand's twelve joints with one of the right hand's five digits, or, perhaps, by counting to twelve with the thumb of one hand and recording multiples of twelve with the digits of the other, it is possible to represent any number from 1 to 60. However it worked, the Mesopotamians' anatomical calculator was a thing of exceptional elegance, and the numbers they counted with it echo through history. It is no coincidence that a clock has twelve hours, an hour has sixty minutes, and a minute has sixty seconds.[27]

THE SUMERIANS' METHOD of finger-counting died with them, but their contemporaries in Egypt and their successors across the ancient world shared a common system that endured for more than three millennia. It is hinted at, in the manner of a Dan Brown conspiracy, by clues scattered at picturesque intervals throughout history and across geographies. Most of these clues are fragmentary and elliptical; one, finally, explains the mystery.[28]

One of the earliest signs of an alternative finger-counting method appeared on the walls of an Egyptian tomb in Ṣaqqārah, a sprawl-ing cemetery located outside the country's ancient capital, Memphis, in the twenty-sixth century BCE. A painted scene shows three men gesturing to one another, hands raised and fingers bent deliberately into different poses, as scribes record their dealings and workers fill buckets with grain. It is clearly a transaction, but an opaque one.[29]

Pliny the Elder, a Roman writer, soldier, and civil servant who lived during the first century CE, is famous for having documented the world around him in precise if not accurate detail.[30] Emblematic of his compendious, rambling approach is the sixteenth chapter of the thirty-fourth volume of his masterwork *Naturalis Historia* (Natural History), in which he recounts the history of statuary in Italy. In passing, Pliny mentions a statue of Janus, the two-faced god, that had been sculpted so that its fingers formed the number of days in the year, "thus denoting that he is the god of time and duration."[31] Some versions of Pliny's words say that Janus signaled 355 days, while others propose 365; either may be true, since the number of days in the Roman calendar year had been adjusted before Pliny's time and the age of the statue is unknown.[32] Whichever is the case, the sculptor must have known of a system capable of counting significantly higher than the Mesopotamian maximum of sixty.

A few decades later in Greece, Plutarch, a biographer of the famous and the doomed, recounted the words of one Orontes in whom the two traits overlapped:

> Orontes, the son-in-law of King Artaxerxes, falling into disgrace and being condemned, said: As arithmeticians count sometimes myriads on their fingers, sometimes units only; in like manner the favorites of kings sometimes can do everything with them, sometimes little or nothing.[33]

In classical antiquity, a *myriad* was a unit of ten thousand, and so Orontes was intimating that mathematicians could count to ten thousand or above using only their fingers.[34] Again, things had clearly moved on from the Mesopotamians' base 60, although Plutarch, like Pliny, was light on the details.

Less equivocal are the *tesserae*. The word has a mysterious ring to it, but it means simply a mosaic tile or other small token. At various times and under various leaders, the Roman state doled out cheap or even free rations of grain to its less-fortunate citizens, where a small, coinlike token called a *tessera frumentaria* represented a ration of a particular quantity.[35] On occasion, the emperor might even issue gifts of *tesserae nummariae*, which, like banknotes, could be redeemed for a sum of money.[36] Found mostly in Egypt, the few extant *tesserae nummariae* show a Roman numeral on one side and, on the other, a hand making a sign that appears to correspond to that same number. They are Rosetta stones for finger-counting, but

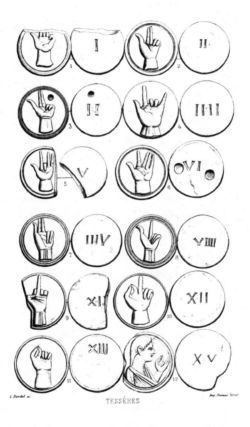

TESSÈRES

○ A gallery of Roman *tesserae* compiled by Wilhelm Froehner, a curator at Paris's Musée du Louvre, and published in 1884. *Wilhelm Froehner, Le comput digital (Macon: Protat frères, 1884), pl. III. https://hdl.handle.net/2027/uiug.30112112048456.*

like the real Rosetta stone, they are incomplete, since none has been found with a number higher than XV, or 15.[37] As such, the insight afforded by these intriguing objects is limited at best, but no matter: one final clue blows the plot wide open.

THE VENERABLE BEDE, as he is often called, was always reluctant to leave the monastery in which he lived. From the age of seven, when his family deposited him at Wearmouth Abbey in the north of England, until his death in 735 CE at nearby Jarrow, Bede spent five decades reading, learning, and, most of all, writing. He wrote about the lives of saints; he commented on passages from the Bible; he wrote homilies to be delivered aloud; and in his most famous work, *Historia ecclesiastica gentis Anglorum* (Ecclesiastical History of the English People), he set down a history of the Church in England that is still read today.[38] He also wrote about *time*.

For Christians, the date of Easter is of paramount importance. The "movable feasts" of Shrove Tuesday, Ash Wednesday, Palm Sunday, Good Friday, and Pentecost, for example, are all defined relative to Easter, so that knowing the date of any one of them means knowing the dates of all the rest.[39] The only problem is that in seeking to know just this, Christianity has repeatedly set petards on which to hoist itself. Although the early Church defined Easter as the fourteenth day of *Nisan*, a lunar month on the Jewish calendar, by 120 CE Christian sects were already bickering over whether to celebrate on the day itself or on the nearest Sunday. Other arguments centered on whether Easter should always be marked after Passover; whether the Jewish calendar was reliable enough in the first place; and whether it mattered if Easter fell before or after the spring equinox. In 325, the Catholic Church laid out the following rules: Easter must always fall on a Sunday; Easter Sunday must be the first one after the fourteenth day of the so-called paschal lunar month; and the paschal lunar month was defined to be the first whose fourteenth day followed the spring equinox.* *Computus paschalis*, the computation of Easter, was both a mathematical and an astronomical problem.[40]

Bede addressed himself to that problem in *De temporum ratione*, (The Reckoning of Time). He digresses from the very first line: "Before discussing the basics of the calculation of time, we have decided to demonstrate a few things, with God's help, about that very useful and easy skill of flexing the fingers."[41] He continues:

> So when you say "one," bend the little finger of the left hand and fix it on the middle of the palm. When you say "two," bend the second from the smallest finger and fix it on the same place. When you say "three," bend the third one in the same way. When you say "four," lift up the little finger again. When you say "five," lift up the second from the smallest in the same way. When you say "six," you lift up the third finger, while only the finger in between, which is called medicus [ring finger], is fixed in the middle of the palm. When you say "seven," place the little finger only (the others being meanwhile raised), on the base of the palm. When you say "eight," put the medicus beside it. When you say "nine," add the middle finger.[42]

In this way, Bede shows it is possible, if not exactly comfortable, to count from zero to nine using only three fingers of one hand. He doubles down on his digital gymnastics by extending his system all the way up to 9,999: the left thumb and index finger manage the tens; the right thumb and index finger record the hundreds; and

---

* Put more simply, Easter almost always fell on the first Sunday after the first full moon after the spring equinox.

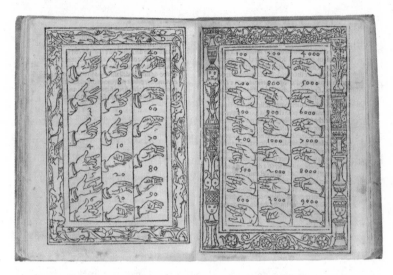

O   An illustrated guide to Bede's finger-counting method, as shown in Filippo Calandri's 1491 math textbook, *De Arithmetica*. *Calandri, Filippo. De Arithmetica. Florence, 1491. Public domain image courtesy of the Metropolitan Museum of Art. https://www.metmuseum.org/art/collection/search/347027.*

finally, the bottom three fingers on the right hand, thousands. These few pages of Latin are the key to three millennia of finger-counting.

Those Egyptians gesticulating about grain some 4,600 years ago? According to Bede's method, the figure on the left is signaling the number 10—or 100, depending on how one interprets left and right within the image. The middleman, so to speak, shows 6 or 6,000, and the rightmost negotiator 7 or 7,000. The system hinted at by Plutarch, Pliny, and other classical writers is likely to be one and the same. The statue of Janus within Rome's forum could easily have indicated 300 on one hand and 65 on the other, while the engraved gestures on the obverse of all those *tesserae nummariae* precisely match those in Bede's *De temporum ratione*.[43] Like the man himself, the system Bede described was venerable indeed.

It is not clear if Bede expected his readers to calculate, as opposed to simply count, with their fingers. Certainly, having explained how to represent numbers with one's hands, Bede pointedly does not describe how to add, subtract, multiply, or divide with them. (While explaining how to compute the date of Easter, he implies that one's fingers should be used only to record intermediate values arrived at via mental arithmetic.)[44] That said, Bede's system does have at least some advantages over Roman numerals, the default notation at the time for writing down numbers. For one thing, there is no easy way to tell what value a particular character in a Roman number represents: consider that 299 is represented as CCXCIX, while 300 is CCC, so that the smaller number requires *more* characters, arranged in a different order, than the larger one. As such, one cannot easily add, subtract, multiply, or divide two corresponding numerals and carry the excess, thus defeating any number of basic arithmetical tricks.

Bede's finger-counting, by contrast, behaves a lot more like our decimal notation. Units are always represented by the same three fingers of the left hand; tens by the two others; the hundreds by two fingers on the right hand; and the thousands by the remaining three. There is never any doubt as to the value held in each "place" in one's hands. Bede might not have wanted his readers to calculate with their hands, but later scholars have constructed increasingly powerful methods for carrying out arithmetical operations using only the hands and the brain.[45]

The tendrils of this ancient style of finger-counting spread across the globe. Bede's account was corroborated by Nicolaus Rhabda of Smyrna, Greece, in the eighth century, while Arabic sources are peppered with oblique references from the seventh century onward. As late as the fifteenth century, math textbooks incorporated instructions on finger-counting as a matter of course, ultimately giving rise to the English word "digit" (and *doigt* in French) to mean numbers less than 10.[46] Think of it: the ancient Egyptians, Plutarch, Pliny, the Venerable Bede, and many more besides, throwing numerical gestures about like philosophical gang signs. Finger-counting prospered beyond their shared system, too, with methods ranging from simple to complex reported across the world. Almost every tribe of people with fingers on which to count has used them to do just that at one point or another.

Finger-counting is not perfect, of course. As the Yupno of Papua New Guinea know, to run out of fingers, or toes, or genitalia, is to run out of numbers. As one Firmicus Maternus noted of the students in his fourth-century classroom, counting with fingers can be fatiguing and difficult to learn. ("*Vides ut primos discentes computos digitos tarda agitatione deflicant?*"—"Do you see how awkwardly beginners in computation bend their fingers?")[47] And as Bede recognized, fingers are not nearly in the same league as the brain when it comes to complex calculations. However high one might be able to count, fingers alone were never going to be good enough.

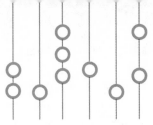

2 THE ABACUS

IF THE HAND WAS the first calculator and the second was a stick, then the third was a rock. The Lebombo bone may be the first known record of a mathematical calculation, but our inbuilt affinity for digital computation means that humans were almost certainly counting on their fingers much deeper in the past. It is likely, too, that as early humans ran out of fingers, or toes, or whatever other body parts they used to count, they would have cast about for a way to extend their counting repertoire. They found one too: without having developed the words for numbers above 10, or 20, or 33, or perhaps for any numbers at all, ancient humans nevertheless found a way to count as high as they were ever likely to need.

THERE IS A thought experiment, often cited in math history books, that starts with a prehistoric shepherd tending her flock. There are, say, forty-two sheep in the flock and she is anxious to know that all of them make it back to her cave after each day's grazing. She cannot count high enough using her fingers or toes, and so uses pebbles to keep count instead. When she shoos the sheep out in the morning, she drops a pebble into a bag for each one. At night, she takes one pebble out of the bag as each sheep enters the cave and places it on a pile. If, after the last sheep has entered the cave, her bag still contains any pebbles, there must be some sheep missing; if, on the other hand, she does not have enough pebbles in the bag to account for all the sheep now in the cave, she must have acquired some new ones, perhaps in the form of a new lamb or a stray from another flock. Without knowing it, the shepherd has invoked the principle of one-for-one correspondence: a physical token of some kind is used as a surrogate for a single object to be counted.[1] Throughout history, we have used the same trick with pebbles, shells, cocoa beans, beads, teeth, and bones; and, yes, with notches on tally sticks.[2]

The same correspondence between counting tokens and abstract numbers may have been the driving force behind Mesopotamia's much-studied system of clay tokens. For four thousand years, from around 7500 BCE onward, the inhabitants of the Near East paired tokens with objects: cones and spheres with barley; cylinders with livestock; tetrahedrons with human labor; and so on.[3] As late as 1500 BCE, a one-to-one correspondence between tokens and objects was still observed. A *bulla* excavated at the site of the ancient city of Nuzi, near Kirkuk in modern Iraq, contained forty-eight tokens representing forty-eight animals: twenty-one ewes, six female lambs, eight rams, four male lambs, six nanny goats, one billy goat, and two female kids.[4]

Yet in earlier times, when the Mesopotamians were still negotiating the transition from counting to writing, the tokens' original one-to-one correspondence had already been tinkered with. Denise Schmandt-Besserat, who theorized that tokens led to writing, noted that small and large cones and spheres seemed to correspond to smaller and larger quantities of barley. The same distinction applied to the small and large tetrahedrons used to record units of work.[5] This is what has been called a size-value system, in which a given symbol or shape represents different values depending on its relative size. A small conical depression meant one *ban* of grain; a larger one meant one *barriga*, a unit six times larger than the *ban*; and so on.[6] From there, some scholars think that the Mesopotamians took a conceptual leap to a *place-value* system, in which a symbol takes on a meaning that is dependent not on its size but rather on its position within a collection of other symbols.[7]

Fortuitously, for such an important part of written math, place-value systems are relatively easy to explain. Take the number 9,999. Each of these 9's means something different to the others: the first represents 9,000, the second 900, the third 90, and the last 9. The *place* in which a given 9 is found gives rise to its *value*. The Mesopotamians and their successors used this principle to create a place-value system for cuneiform numbers. Their system represented fractional values, too, so that the Sumerians were able to compute pi to two decimal places, the square root of 2 to five decimal places,* and reciprocals (that is, 1 divided by another number) for values in the billions.[8] But the Sumerians may have done far more than invent place-value notation. There are hints in ancient Sumer of an entirely new kind of mathematical tool. There are hints of a calculator.

There is a collection of cuneiform tablets—the so-called Old Babylonian Lu Series—that dates to the early part of the second millennium BCE and that forms part of a long tradition of "lexical lists." These are dictionaries that translate words from Sumerian, as spoken by the inventors of cuneiform, into the Akkadian

---

* The Sumerians' computed value was 1.414222; the true value is 1.414214[. . .].

language of their successors.[9] The Lu Series enumerates professions, and one word in particular, *šid* (𒐲), pronounced "shid" and standing for "count," "counting," or "number," occurs in a number of different job titles.[10] With the *lu* of the tablets' title meaning "man," these occupations have been interpreted to mean, variously, "man of the counting token," "man of the wooden counting board," and, most hopeful of all, "man of the abacus."[11] These gnomic entries are uncorroborated by written evidence or physical artifacts, but the fact that they exist at all suggests that Sumerians may have invented the first ever purpose-built calculating device. The *abacus*.

Before wading into the disputed murk that surrounds the birth of the abacus, it is necessary to describe the star of this chapter slightly ahead of time. Now, you may already know about the abacus. Like a penny-farthing or a pitchfork, the abacus is, to many people, a familiar but redundant artifact: you may have owned a toy abacus as a child, or inherited one from a family member, or simply absorbed the idea of the abacus through cultural osmosis, even if you have never used one to carry out a calculation. Nevertheless, a brief introduction will help anchor some of what follows.

Most abacuses take the form of a rectangular frame crossed by a series of parallel wires or rods. Each wire is strung with beads that can be slid along its length. Wires may run horizontally or vertically, but each one typically carries the same number of beads. In some variants, a bar runs across the wires and divides the abacus into two sections.

Abacuses are used to represent place-value numbers. Each of an abacus's wires corresponds to a single digit, so that the number 32 is represented on a decimal abacus by sliding 2 beads away from the others on the "units" wire and 3 beads on the "tens" wire, like this:

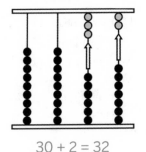

O Setting a simple decimal abacus to represent the number 32. The leftmost wire represents thousands and the rightmost wire represents units. Gray beads are those that have changed position; black beads are unchanged.

30 + 2 = 32

To add two numbers together—say, to add 29 to our initial value of 32—we work from right to left, adding units to units, tens to tens, and so on, carrying any excess to the next column as we go. Experienced abacists tend to sum each column in their head and then move all the affected beads at once. We will do the same here.

First, we add the units. Our abacus has 2 beads set on the units wire, and we want to add 9. Our abacus only has 9 beads on each wire, this being a decimal abacus, and so we cannot simply add 9 beads to the units wire—with 2 beads already set, there are only 7 remaining. Instead, we add 1 bead to the tens wire and *subtract* 1 bead from the units wire, since 10 − 1 = 9. As a result, we see that adding 9 to our initial value of 32 gives us 41:

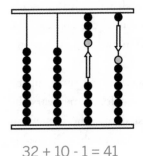

O Adding 9 to 32 in two steps: add 10, subtract 1. The result is 41.

32 + 10 - 1 = 41

Finally, we move on to the tens. We add 20, or 2 beads on the tens wire, to give us 6 beads on the tens wire and 1 bead on the units wire. We have computed 29 + 32 = 61, all without a pen, pencil, or calculator.

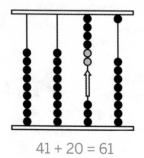

41 + 20 = 61

O   Adding 20 to 41 in a single step to give 61.

Subtraction is similar, except that we start with the most significant wire, not the least. And with these two basic operations in hand, a skilled abacist can multiply and divide, and, if necessary, carry out even more complicated operations.

GIVEN THE LACK OF concrete evidence, why do some historians think that the Sumerians invented the abacus? The answers lie on a spectrum between speculation and divination, and some of them might just be correct.

At the "wishful thinking" end of the spectrum is David Eugene Smith, whose 1923 *History of Mathematics* was for many years the standard text on the subject. Smith suggested that the cuneiform character *šid*, as seen in the Lu tablets' job titles, might conceivably be shaped like an abacus.[12] It is true that many cuneiform symbols are descended from pictures of actual things, but Smith's theory suffers from an absence of supporting evidence and, more acutely, from the fact that a substantial minority of cuneiform symbols look a lot *more* like an abacus than 𝅘𝅥 does.[13]

Elsewhere, Denise Schmandt-Besserat herself wondered in her early writings if certain "perforated" tokens might have been strung onto an abacus in some way.[14] Again, a dearth of physical evidence makes this unlikely.[15]

More recently, Georges Ifrah, the author of another authoritative guide to the history of mathematics, has written that the Sumerians "must" have used a counting aid of some kind. (Ettore Carruccio, a prominent Italian mathematician, makes a similar assertion.[16]) Taking note of the clay tokens that have been excavated across the Near East, and observing that some present-day cultures lay pebbles in rows to count large numbers, Ifrah contends that the Sumerians could not have computed their square roots, multiplication tables, and reciprocals without first coming up with a similar mechanism. Perhaps, he says, the Sumerians could have scratched some lines in the dusty Mesopotamian earth and placed their tokens on them as an impromptu abacus? No wire, no wood, no beads; just clay and soil.[17] No traces of such a practice have been found, but Ifrah goes on to raise a more intriguing hypothesis regarding the Sumerians' successors, the Akkadians.

Toward the end of the twenty-fourth century BCE, a king named Sargon conquered many Sumerian city-states, eventually reaching the Persian Gulf, in whose briny waters he symbolically washed his bloody weapons. To anchor his new realm, Sargon built a city that he named Agade, or Akkad, which in turn gave its name to the kingdom of Sumer and Akkad.[18] (Coincidentally, it also gave Sargon of Akkad possibly the most luridly fantastical sobriquet of any ancient ruler.) The Akkadians spoke their own language but wrote it with the Sumerians' cuneiform script, necessitating Sumerian-to-Akkadian dictionaries such as the Lu tablets.[19]

The Akkadians also inherited the Sumerians' base-60 numerals but, over time, replaced them with a simpler base-10 system. Decimals had been lurking in Sumerian numbering for some time: the intervals from 1 to 60 and then to 3,600 were unwieldy in everyday use, so the Sumerians often used multiples of 10 to make things easier. Thus, a mathematically literate Sumerian would have counted in terms of 1, 10, 60, 600, 3,600, 36,000 and so on. All the Akkadians did was remove the ambiguity from this hybrid notation. Georges Ifrah connects the resulting decimal system to an artifact discovered in the ruins of Susa, an ancient city in what is now Iran. It is a brick inscribed with a grid of squares, ten on each side, that Ifrah asks us to think of as an abacus on whose columns tokens or pebbles could have been placed to represent the digits of a decimal number.[20]

Whether Ifrah's brick is an abacus or not, there is circumstantial evidence that students in Babylonia, a state formed after the fall of the Akkadian Empire, used a calculating device that shared some of the abacus's characteristics.[21] The case is taken up by one Jens Høyrup, a Danish historian specializing in Mesopotamian mathematics, who presents a cuneiform tablet on which the scribe explains how to square the number 650. The correct answer is 422,500, but the writer mistakenly gives 424,000. Or, in base-60, the scribe arrives at 1,57,46,40 rather than the correct value of 1,57,21,40. Even in cumbersome sexagesimal notation, the error is an obvious and very specific one: the writer has mistakenly added 25 to the third digit of their final value.

From other tablets, we know that the Babylonians used the same "partial products" method that is taught in elementary schools today to carry out large multiplications, and Høyrup detects the fingerprints of just such a method in this error. The errant number, 25, crops up when one tries to decompose the sexagesimal form of 650 × 650 into its constituent parts, and this is where things may have unraveled. The scribe accidentally added 25 to the third digit of their running total twice, not once. Had they been tracking intermediate results on a clay tablet, they would have noticed their error. But if they were using a device that could record only a single number at once—say, a clutch of clay tokens and wooden tablet inscribed with vertical lines—then the error would be an easy one to make. *Did I just add 25 to the third column? I don't think so. I'll add it now.*

Høyrup also draws attention to a similar error on a different tablet, which suggests that a value from a partial product was added to the wrong column. Again, it is a mistake that would have been easy to detect if the writer had recorded each step of their work, but when a calculating tool has no memory, like an abacus, the onus is on its user to do the remembering.[22]

None of this is conclusive. But the ancient world had a habit of neglecting its most consequential inventions: the origins of writing, papyrus, coins, printing, and a host of other tools that revolutionized intellectual and bureaucratic life have all been lost to time. All we can say is that the abacus *may* have been invented between three and five millennia ago in the land between the Persian Gulf and the Mediterranean Sea. There is a lot of room in there for the truth to hide.

THE HISTORY PROPER of the abacus starts in Greece.

In 1846, one Alexandros Rizos Rangavis wrote to Jean-Antoine Letronne, a prominent French archaeologist, to ask for his opinion of a marble tablet that Rangavis had excavated on the island of Salamis in the Saronic Gulf, just across the water from Athens.[23] Rangavis was the sort of globetrotting nineteenth-century aristocrat through whose pen history was recorded: born to Greek parents in Constantinople, raised in Bucharest and Odessa, and having served in the Bavarian Army, Rangavis eventually settled in Greece, where he helped found the country's first modern university. He later became Greece's first foreign minister, and in 1867 and 1868, he served as the country's first ambassador to the United States. (He liked American women; he disliked American manners.)[24] But today,

Rangavis's name has been eclipsed by that one find on Salamis: a white marble slab almost five feet in length and half that in width, crossed by eleven inscribed lines and marked with Greek numerals along three edges. Dating to around 300 BCE, it is likely to be the world's earliest known abacus.[25] Or rather, as abacuses used with loose tokens are called, the world's earliest *counting board*.

The ancient Greeks were the first to leave palpable traces of their use of counting boards. Towards the end of the fifth century BCE, for example, the Greek historian Herodotus wrote that the Egyptians "manipulated the pebbles (*psephoi*)" from right to left, in the opposite direction to the Greeks.[26] Scholars agree that Herodotus was describing the counting tokens that the Egyptians* and the Greeks used on their respective counting boards.[27] Allusions to *psephoi* were plentiful elsewhere in Greek art and literature. In a speech known as "On the Crown," delivered in 330 BCE, the famed politician Demosthenes† invokes counting tokens over and over in metaphors too tortuous to repeat.[28] At around the same time, in Apulia, Italy's "heel," a Greek settler was putting the finishing touches to the so-called Darius Vase, a colossal ceramic vessel emblazoned with images of the Greek gods, the Persian king Darius, and, lowering the tone a little, a tax collector adding up tributes from conquered lands. In front of him, very plainly, is a table scattered with white dots and inscribed with Greek numerals—*psephoi* on a counting board.[29]

The Salamis tablet was the imposing real-life double of the portable counting board depicted on the Darius Vase. Its eleven transverse lines are assumed to correspond to the eleven characters also

---

* Evidence for Egyptian abacuses and counting boards is exceptionally scarce. Among the few possible candidates is a "dot diagram" sketched out on a papyrus calendar dating to around 1225. It may represent a set of pebbles on a counting board, or, equally, it may be the doodle of a bored scribe.

† Demosthenes famously cured a childhood speech impediment by practicing speaking with a mouthful of pebbles, or *psephoi*.

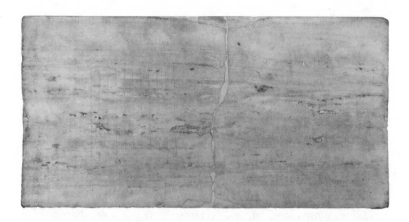

O The earliest known abacus, or counting board, unearthed on the Greek island of Salamis and dating to the fourth or third century BCE. The tablet measures around five feet long and two and a half feet wide. *EM 11515: Epigraphic Museum, Athens © Hellenic Ministry of Culture and Sports / Hellenic Organization of Cultural Resources Development (H.O.C.R.E.D). Photograph by Giorgios Vdokakis.*

inscribed on its surface: XⲤHⲤⲀⲅHⲤTX, repeated three times on three different edges.[30] Read from right to left, the tablet's legend suggests that it was meant for accounting: X represents one-eighth of an *obol*, a Greek coin; T is one-quarter, ⲤС is one-half, and I is one full *obol*. Ⲅ is one *drachma*, a coin worth six *obols*.[31] After that, the tablet switches from currency to plain old numbers. The Greeks did not have dedicated numerals, as we do, but used letters instead: Ⲅ (capital *pi*) stands for the word *pente*, or 5; Δ (capital *delta*) stands in for *deka*, or 10; H (capital *eta*) is *ekato*, or 100; and X (capital *chi*) represents *chilia*, or 1,000. Rounding out the legend are a pair of composite symbols: Ⲅ means Ⲅ × Δ, or rather 5 × 10 = 50; while Ⲅ stands for Ⲅ × H, or 5 × 100 = 500.[32]

Although the Salamis tablet was labeled for monetary calculations, it would have made a fine general-purpose counting board for any user with the presence of mind to treat its currency symbols as abstract numbers.[33] In fact, analysis of errors in some of the calculations sprinkled throughout Herodotus's *Histories* showed that he probably used a Salamis-style counting board.[34] All in all, there is convincing evidence that the counting board was part of the fabric of ancient Greek culture—a tool to reckon with, both literally and metaphorically; to be invoked by politicians and memorialized on commemorative vases. Why, then, are there so few of them? The linchpin work on classical counting boards, Mabel Lang's 1957 *Herodotus and the Abacus*, lists only thirteen known examples, and the archaeological world has not exactly been rocked by revelations of new counting boards since then.[35]

There are two theories. Boards made of perishable wood may have been more common, and certainly more portable, than durable marble or stone boards such as the Salamis tablet. More intriguingly, it may have been that counting boards were simply *imagined* whenever they were needed. A counting board is not a complex device, and it would have been easy to visualize a set of lines on any reasonably flat surface.[36] Who needs a counting board when you have a table, a rock, or the stairs leading to the Temple of Apollo?[37]

THE ROMANS MAY NOT HAVE invented counting with tokens, but they left an indelible linguistic imprint on it. The Greek term *psephoi* gave way to the Latin *calculi*, from *calx*, meaning "pebble" or "gravel." That, in turn, evolved into the verb *calculare*, "to calculate." *Calculones* were enslaved people who taught arithmetic; *calculatores* were citizens who did the same.[38] (By extension, all devices for calculation would eventually become "calculators.") For the counting board itself, the Romans took a Greek word, *abax*, and turned it into the Latin *abacus*, and the rest, as they say, is history.[39] The origins of the word *abax*, however, are less clear-cut.

There is a persistent yarn that says the Greek term *abax* comes from *abaq*, an old Hebrew word meaning "dust," and that, once upon a time, a denizen of the Near East in need of a temporary calculator would naturally have scratched the lines of a counting board on the dusty ground. J. M. Pullan, for instance, author of one of the few book-length histories of the abacus, says that the practice of counting with pebbles would have become inextricably associated with the act of hunkering down to sweep aside some sand or dust and then casting one's pebbles onto the ground. In this way, *abaq* (and later *abax*) may have referred to the ground on which the first pebble calculations were done.[40]

As neat as all of this sounds, the evolutionary line from *abaq*, "dust," to *abacus*, "counting board," is tenuous at best. Both the Greeks and the Romans used *abax* and *abacus* more generally to refer to all manner of flat objects: the tops of architectural columns, plastered wall panels, gaming boards, sideboards, and wooden boards for kneading dough.[41] *Abax* and *abacus* were by no means reserved for calculating devices. Even more incongruous is that the counting boards used in Greece and Rome were *not* abacuses, at least not in the modern sense of the word. The word *abacus* is Latin but the device that we call an "abacus" is not. To find the true origin of the abacus, we need to travel east from one empire to another.

A S GREECE AND ROME WERE coming to grips with the counting board, China was going through serious growing pains. Two and a half centuries of strife between seven warring kingdoms, punctuated by endless battles and political intrigue, had come to an end in 221 BCE only when Qín Shǐ Huáng,* the self-titled first emperor, unified China by force.[42] Not that life under Qín's iron

---

* Pronounced "Chin," Qín may have given China its name.

fist was much better than it had been during the so-called Warring States period. In 213 BCE, for instance, the new regime systematically burned books of history, poetry, and literature that conflicted with the Qín worldview—but, tellingly, books related to medicine, farming, or science were spared.[43]

The Qín dynasty was ousted fifteen years after it began, and, in the four centuries of the Han era that followed, the arts and sciences were feted even as the government maintained a tight grip on the new state of China.[44] As a result, the Han period accounted for at least two of ancient China's "four great inventions": cheap, portable paper came to prominence during this time, supplanting heavy bamboo and costly silk as the preferred medium for writing, while the newly discovered magnetic compass was put to use in divination and *fēng shuǐ*.[45] (Gunpowder would follow in the ninth century, printing in the eleventh.)[46] It was against this backdrop that, around 190 CE, one Xu Yue wrote a book called *Shù Shù Jì Yí* (數術記遺), or *Notes on Traditions of Arithmetic Methods*. It is this book in which the abacus as we know it was first revealed.[47] Probably.

Xu Yue was an astronomer and mathematician, but little else is known about him. And although *Notes* is credited to Xu, there is a possibility that a later author, Zhēn Luán, reputed to have written a substantial commentary on the original text, may have written the book in its entirety. Either way, attaching his name to that of Xu, a famous mathematician, was a savvy move on Zhēn's part. *Notes* became required reading for China's civil service exam in the seventh century, elevating Zhēn's name alongside Xu's.[48] And yet, *Notes* is an odd book, ranging far and wide into apparently unrelated mathematical miscellany. Among other things, it describes magic squares (an ancient relative of the sudoku puzzle); three separate notations for representing very large numbers (which may have been a comment on Daoist notions of infinity); and two different physical systems of calculation. One was old, one was new, and both were vital to the Chinese way of life.[49]

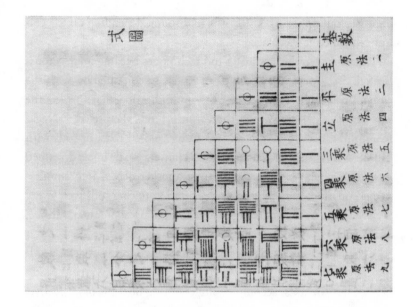

○ Chinese counting rod numerals. There are few surviving printed examples of Chinese rod numerals, but the technique also made its way to Japan, where this table was published in 1712. In the West, this pyramid of numbers is called "Pascal's triangle": each of the numbers along the pyramid's edges is 1, and each interior number is the sum of the two numbers above it and above and to its right. The zero-like figures represent just that—zero. 関孝和 *[Seki, Takakazu]*, 括要算法 *[Katsuyō Sanpō], vol. 1* (升屋五郎右衛門 *[Goroemon, Masuya], 1712), 27, https://doi. org/10.11501/3508173. Image courtesy of the National Diet Library.*

The first of *Shù Shù Jì Yí*'s calculation aids came in the form of short rods called *suàn* (算) or *chóu* (籌), which were placed on a grid to carry out mathematical operations.[50] Each digit was made from one or more rods so that, from 1 to 10, they looked like this: _, =, ≡, ≣, ≣, ⊥, ⊥, ⊥, and ≣. Cleverly, digits 1 to 5, each consisting of the corresponding number of rods, could be added to one another directly—and then, whenever a numeral ended up with more than five horizontal rods, those five rods were exchanged for a single vertical one. When placed on the accompanying grid, each individual digit was multiplied by 1, 10, 100, and so on, depending on the column in which it was placed.[51]

As time went by, counting-rod techniques became more sophisticated. *Suàn* were increasingly put to use in combination with the "nine-nine" table, a mnemonic for multiplication that formed the software to the hardware of the counting rods themselves.[52] Negative numbers could be represented using colored rods, or rods with different cross-sectional shapes. One Han-era math book described how to represent fractional numbers; another delved into simultaneous equations and cube roots.[53] China's counting rods were every bit as capable as the counting boards used in Mesopotamia, Greece, and Rome.

The earliest archaeological evidence for counting rods are rod-like numerals shown on coins minted during the Warring States period, but the rods' accompanying counting board is more elusive. Some texts say that a special carpet should be laid out before calculating; others suggest using a table or even the ground itself. Even more so than its Western counterpart, the Chinese counting board may have been more of a conceptual aid than a physical one.[54] At any rate, counting rod digits were sufficiently useful to give rise to written numerals that persisted in one form or another until at least the sixteenth century.[55]

The curious thing is that counting rods largely disappeared from China's cultural memory at around the same time as their written offspring.[56] The reason can be traced all the way back to the *Shù Shù Jì Yí* and the *other* mechanical counting system it describes.

N THE THIRD VOLUME of his eye-poppingly comprehensive series of books on *Science and Civilisation in China*, Joseph Needham CH, FRS, FBA, Marxist, biochemist, historian, and Sinologist extraordinaire, translated part of the *Shù Shù Jì Yí* into English for the first time.[57] His translation reads as follows:

> The ball-arithmetic method holds and threads together the Four Seasons, and fixes the Three Powers (heaven, earth and man) like the warp and weft of a fabric.

Not the most illuminating sentence, but Zhēn Luán's later commentary makes things much clearer:

> A board is carved with three horizontal divisions, the upper one and the lower one for suspending the travelling balls, and the middle one for fixing the digit. Each digit (column) has five balls. The color of the ball in the upper division is different from the color of the four in the lower ones. The upper one corresponds to five units, and each of the four lower balls corresponds to one unit. Because of the way in which the four balls are led (to and fro) it is called "holding and threading together the Four Seasons." Because there are three divisions among which the balls travel, so it is called "fixing the Three Powers like the warp and weft of a fabric."

In writing this, Zhēn had used a word—*dài*, meaning "belt" or "ribbon"—that led Needham to think that the five balls in each column must have been threaded onto a string or something similar.[58] Put simply, Needham wrote, Xu and Zhēn must have been describing an abacus: the iconic Chinese *suàn pán*, or bead framed abacus, with its columnar beads divided into lower and upper registers called earth and heaven, where the lower beads represent one unit each and the upper beads, five.[59]

Granted, there was what might be called a decent interval between the *Shù Shù Jì Yí* of 190 CE and the earliest detailed account of the *suàn pán*, which arrived some thirteen centuries later in a book entitled *Suàn Fǎ Tǒng Zōng* (Systematic Treatise on Arithmetic). The author of that later book, Chéng Dà-Wèi, describes an instrument that had acquired an extra bead in both the "heaven" and "earth" portions of each wire, but which was functionally identical to the abacuses still used today by diehard adherents both in China and in the Chinese diaspora around the world.[60]

The *suàn pán* was very much an evolution of what had gone before. The separation of each column into units and fives recalled the structure of counting-rod numerals, while many of the rules for

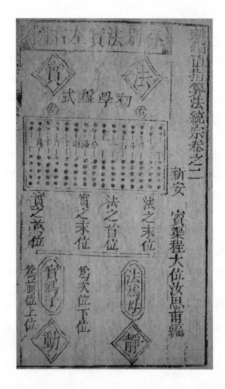

O The earliest known depiction of a Chinese abacus, taken from Chéng Dà-Wèi's *Suàn Fǎ Tǒng Zōng* (Systematic Treatise on Arithmetic) of 1593. *Dawei Cheng, Suan Fa Tong Zong Da Quan, vol. 1, n.d., 39, http://echo.mpiwg-berlin.mpg. de/MPIWG:OZHYTPUF. CC BY-SA 3.0 DE image copyright Max Planck Institute for the History of Science.*

manipulating the *suàn pán*'s beads were similar or identical to those used with rods.[61] The *suàn pán* mimicked the counting rods' imaginary grid, too, in that it assigned no fixed values to its columns, leaving the user to pick the units they needed. As such, it was as capable of handling very large numbers as it was very small ones, or of reserving some columns for intermediate results during complex calculations—a trick not easily replicated on Western counting boards with their fixed monetary units.[62] Even the terminology of the *suàn pán* was replete with callbacks to its past. The word *suàn*, once used for counting rods, now had the more general meaning of "to count" or "to calculate." (*Pán* stood for "tray" or "table.") The character for *suàn* (算) hid another historical clue, too, in that it contained the so-called bamboo radical (竹 or ⺮), or graphical component, since most counting rods had been made from bamboo.[63]

B Y   D E G R E E S the *suàn pán* made itself felt in the wider world, reaching Korea around 1400 CE and then Japan no more than two centuries later.[64] Japan in particular took to the abacus as wholeheartedly as its Chinese inventors, although exactly when they did so is a matter of debate. Japan's upper classes disdained base pursuits such as making or counting money and so, although the Japanese abacus, or *soroban*, is first mentioned by 1600, it may have been in common use before then.[65] Given that Japanese travelers visited China as early as the seventh century, and that some prominent Chinese figures were reputed to have immigrated to Japan soon after that, it may have been much, much earlier.[66]

By the twentieth century, the *soroban* was ubiquitous. Not only that, but in moving from seven beads per wire to five, one in "heaven" and four on "earth," it had regained the original form of the abacus first described by Xu Yue and Zhēn Luán almost two thousand years before.[67] It was used by shopkeepers, restaurateurs, insurance adjusters, bank clerks, and accountants.[68] There were even regular abacus competitions, with no fewer than forty thousand contestants vying in 1942 for Japan's top prize.[69]

This love affair with the abacus culminated in 1946, in an extraordinary moment that positively dripped with symbolism. East versus West; old versus new; hubris versus humility; tradition versus progress; you name it, the exhibition match between Kiyoshi "The Hands" Matsuzake of the Japanese Ministry of Postal Administration and Pvt. Tom Wood of the U.S. Army had it all. But this was no boxing match. Instead, in an event sponsored by *Stars and Stripes*, the U.S. Army newspaper, Matsuzake wielded his *soroban* and Wood took the helm of a state-of-the-art electric calculator to face off in a contest of mathematical speed and accuracy.[70] It turned out to be no contest at all. Matsuzake won at addition, subtraction, and division, while Wood bested him only at multiplication. With echoes of China's ancient nine-nine table, Matsuzake's success was attributed to the way he carried out simple calculations in his head rather than on his abacus.[71]

The *soroban*, which manifested on a spectrum from desk-spanning behemoth to portable-but-not-pocketable minnow, continued in rude health until the very end of the twentieth century. Abacus arithmetic was taught in Japanese elementary schools until the 1970s, and, as late as 2019, around forty-three thousand people were still taking lessons at one of six thousand private *soroban* schools.[72] Interest has slackened in recent times, but the two-thousand-year story of the *suàn pán* and *soroban* is not over yet.

THERE IS, UNFORTUNATELY, a fly in this ointment. The conventional view of the abacus as emanating from the mysterious Orient, invented in closeted China and taken up in traditionalist Japan, may have a Western origin story of its own.

O A modern replica of the classical Roman hand abacus, made by Prof. Dr. Jörn Lütjens of Hamburg University. The final column, speculatively divided into three sections, may have been used to represent fractional numbers: halves, thirds, and quarters. © *Jörn Lütjens https://joernluetjens.de/sammlungen/abakus/ abakus.htm.*

One is in the Bibliothèque nationale in Paris, one in Rome's Museo Nazionale Romano, and one in a museum in the Italian Alps. Attested by only these three surviving examples, the device known as the Roman hand abacus is even less well understood than its ancient Chinese counterpart.[73] If the history of the *suàn pán* comes mostly from written accounts, the Roman hand abacus is the opposite: there are no surviving texts to explain how or when the hand abacus* was invented or whether it was ever used as a serious mathematical tool.[74]

---

* There is yet another version of the abacus, or counting board, that suffers from the same problem as the Roman hand abacus. In 1590, a Spanish priest named José de Acosta wrote of an intriguing device used in the Inca Empire:

In order to effect a very difficult computation for which an able calculator would require pen and ink [...] these Indians make use of their kernels of grain. They place one here, three somewhere else and eight I know not where.

That said, it is possible to imagine how the Roman hand abacus might have come about. At some point or another, some nameless Roman decided to unite counting tokens and counting board into a single bronze device, where sliding buttons ran in slots to simulate the columns of the counting board. In the tradition of the counting board, each of the first seven columns was labeled with a Roman numeral: I for ones, X for tens, C for hundreds, ⅭⅠↃ (or "ⅭⅠↃ") for thousands, ⅭⅭⅠↃↃ for tens of thousands, ⅭⅭⅭⅠↃↃↃ for hundreds of thousands, and, lastly, X̄ for millions. Like the *suàn pán*, each of these decimal slots was broken into a lower portion containing four buttons and an upper portion with a single button.

Diverging from the Chinese model, two non-decimal columns were also present. The first, labeled O, contains five beads in the lower portion and one above, so that it can represent the numbers zero to eleven. It was likely used for ounces, or twelfths. The second non-decimal column is unlabeled, and frankly, no one is entirely certain what it is for.[75]

All of this leads to a situation in which no one knows which abacus inspired the other or, indeed, whether the two could have grown up independently. If Xu Yue's account is to be believed, the Chinese were using bead frame abacuses no later than 190 CE—but then, Zhēn Luán's more convincing commentary did not arrive until four hundred years later. On the other hand, none of the handful

---

They move one kernel here and three there and the fact is that they are able to complete their computation without making the smallest mistake. As a matter of fact, they are better at calculating what each one is due to pay or give than we should be with pen and ink.

This was the *yupana*, a kind of counting board used with kernels of grain rather than pebbles. The *yupana* is a thoroughgoing mystery: If any contemporary accounts of its rules or its history have survived the Incas' wars with their Spanish colonizers, their contents remain locked behind the Incas' undeciphered knotted-rope writing.

of surviving Roman hand abacuses has been properly dated, which makes it impossible to assign precedence. From both temporal and geographical points of view, everything is still up for grabs.

What is plainer to see is that the Roman hand abacus disappeared without a trace. As China wholeheartedly embraced the bead framed abacus, the Romans, and Europe as a whole, were a little too preoccupied to be embracing much at all.

T O S AY T H AT Europe fell into darkness as a host of barbarian tribes ravaged Rome, the eternal city, is to put too bleak a spin on things. Certainly, the empire had seen better days, divided as it was into Western and Eastern halves: the Eastern empire, with its capital at Constantinople, or modern-day Istanbul, was doing just fine, thanks; but for the Western empire, still clinging to the city of Rome and its ancestral possessions in Europe, the fifth century CE was one long, drawn-out humiliation. Rome was sacked in 410 and then again in 455 before the teenage emperor Romulus Augustulus was finally deposed in 476 by the Germanic troops of a general named Odoacer.

For the most part, life was not markedly different under the Germanic tribes that replaced Rome's old elite. The incomers were drawn to Christianity, by then Rome's state religion, and picked up Latin to help govern their new lands.[76] When the storied King Charlemagne was crowned in the year 800, ushering in a minor renaissance of the intellectual arts, Pope Leo III named him "most serene Augustus" and emperor of the "Holy Roman Empire."[77] Everything old was new again.

Two things did *not* flourish under the new regime: neither the counting board nor the Roman hand abacus were anywhere to be seen. For the hand abacus, this was the end of the line. But for the counting board, the fall of Rome was an inconvenience and the rise of the Holy Roman Empire an irrelevance. In 1000 CE or thereabouts, the counting board made a blockbusting comeback, courtesy of none other than the new pope, Sylvester II.

Born around 950 CE in Belliac, France, the man who would be pope migrated from France to Spain and on to Italy over the next half century, climbing as he did so the rungs of the Catholic ladder from monk to abbot to archbishop and, finally, to the papacy itself. Always on the scientific end of the papal spectrum, when Sylvester, né Gerbert, published a pair of books describing how to use what he called an abacus, the clout that comes with being the pope ensured that they were widely read.[78] Gerbert somehow neglected to depict the device itself, but a text penned by a student of his rectified that omission so that we know that Gerbert's "abacus" took the form of a counting board with twenty-seven decimal columns. Furthermore, we know that it was used with tokens inscribed with newfangled "Arabic" numbers—the numerals we still use today, most likely taught to Gerbert in Barcelona, hard up against the border of al-Andalus, or Muslim-controlled Spain.[79]

It was the start of a long second life for the counting board. The first dedicated metal counting tokens were minted in Italy around 1200, and soon boards and counters were everywhere in Europe.[80] Echoing Demosthenes, Martin Luther used counters as a metaphor in the sermons he delivered:

> To the counting master all counters are equal, and their worth depends on where he places them. Just so are men equal before God, but they are unequal according to the station which God has placed them.

Shakespeare called on them in *Julius Caesar*, implying that Brutus is so greedy as to steal even worthless counting tokens:

> When Marcus Brutus grows so covetous
> to lock such rascal counters from his friends be ready,
> gods . . .

And as late as the nineteenth century, Johann Wolfgang von Goethe could write of them in *Faust*, relying on his audience to know that *Rechenpfennige* (counting tokens) were inherently valueless:

> Did you think they'd give you real money and goods?
> In this game even worthless counters
> Are far too good for you.[81]

Shakespeare, Luther, and Goethe were right: counters had no numeric or monetary value, except when placed in their natural habitat of the counting board. But there was no reason why the humble counting token could not become a vehicle for values of a more abstract kind. In France, counters were often stamped with the emblems of their owners: the crown, the royal mint, the treasury, the landed gentry, and so on.[82] In the Spanish-occupied Netherlands, the Dutch issued counters bearing a motto designed to rouse the populace: "Justice is struck dead, the truth lies in distress." Their occupiers played the same trick, minting tokens that carried the likeness of the Spanish king and warned, "Woe unto the people which rises against my house."[83]

Equally, there was nothing to stop tokens being imbued with value of a baser kind. In an entirely characteristic gesture, Louis XIV, the "Sun King" of France and ringmaster of proceedings at the Palace of Versailles, gave sets of gold and silver tokens to favored officials as New Year gifts. Many of these were subsequently melted down to recover their constituent metals, thereby cementing the irony of a valueless counter made of precious metal.[84]

○——○

THE RESURRECTED COUNTING BOARD had a good run. In Britain, for example, a book on arithmetic first printed in 1543 maintained a chapter on the counting board through forty-five separate editions before finally going out of print in 1700.[85] Many other European books addressed the same subject.[86] In the end, the thing that did for the counting board was something well known to the architect of its revival, Pope Sylvester II. And, adding insult to injury, China, whose bead framed abacus would handily outlast the counting board, was an unwitting accessory to the crime.

The symbols we call Hindu-Arabic numerals—0, 1, 2, 3, and so on up to 9—had a long gestation. Their earliest known ancestors were first used in India in the third century BCE and were called the Brahmi numerals.[87] Like Greek and Roman numerals, these lacked both a place-value system and the concept of zero; unlike their classical counterparts, they went on to acquire them by the first and the fourth centuries respectively. Arabic scholars brought Indian numerals to Spain in the eighth century, where, finally, Europeans such as the future Pope Sylvester II could learn of their existence.[88]

Hindu-Arabic numerals bridged the gap between writing and computation. Roman numerals were easy to write but painful to compute with, while the counting board could compute but not record. With a pen, ink, and paper, it was now possible to calculate *and* to keep records at the same time. Two scholars did more than any others to popularize this new way of thinking: Muḥammad ibn Mūsā al-Khwārizmī of Baghdad and one Leonardo of Pisa. Both would acquire more famous nicknames before long.

Born late in the eighth century, al-Khwārizmī studied at the Bayt al-Ḥikmah, or House of Wisdom, in Baghdad, then the teeming, cosmopolitan capital of a caliphate that stretched to Europe, North Africa, and Asia.[89] Early in the ninth century, al-Khwārizmī wrote a pair of books that would cement his place in the mathematical canon: *Algoritmi de numero Indorum* (Al-Khwārizmī Concerning the Hindu Art of Reckoning) and *Al-Kitāb al-mukhtaṣar fī ḥisāb al-jabr wa'l-muqābala* (The Compendious Book on Calculation by Completion and Balancing). It is hard to overstate the impact of these two works. The first, which survives only as a Latin translation, gives a guided tour of the Indian number system. It also gave al-Khwārizmī his Latinized nickname of "Algoritmi," from which, in turn, comes the modern term "algorithm"—a mathematical recipe that tells the cook how to solve a complex mathematical problem by following a series of simpler steps. Al-Khwārizmī's second book describes how to deal with equations containing unknown values—the x's and y's that give algebra its power—and it also gave algebra its name through the word *al-jabr* in its title.[90]

The other book most closely associated with Arabic numerals and pen-and-ink reckoning came from the West, but despite its name, Leonardo of Pisa's *Liber abaci* of 1202 was not concerned with the counting board. Leonardo had been exposed to many different flavors of mathematics on a tour of Egypt, Syria, Greece, Sicily, and Provence—all of which he rejected in favor of *modi Indorum*, the Indian system.[91] *Liber abaci*, then, explained how one might avoid the counting board by means of pen, ink, and Indian numerals, and it ranged across arithmetic, algebra, and geometry along the way. Almost coincidentally, in posing a problem regarding rabbit populations, Leonardo—later called Fibonacci, probably after an old family surname—inadvertently stamped his name in the annals of mathematical history. Today, he stands with Al-Khwārizmī as one of the progenitors of the way we count and calculate.

From the beginning, there was bad blood between the "algorists" and the "abacists." In Italy, in Fibonacci's time, merchants embraced the new numerals, developing double-entry bookkeeping as they did so to make even better use of them. But the forms of the new numbers were still fluid, and fraud was rife. Perhaps as a result, in 1299 the city of Florence forbade the use of Hindu-Arabic numerals in banks. A later edict in Padua stated that book prices had to be listed in Roman numerals.[92] Outside Italy, Roman numerals clung on in

mercantile circles until the fifteenth century and, in more conservative environs such as monasteries, even longer than that.[93] Even in
1543, Nicolaus Copernicus's *De revolutionibus orbium coelestium* (On
the Revolutions of the Heavenly Spheres), a text widely considered
to mark the start of the Scientific Revolution, contained a mix of
Roman, Hindu-Arabic, and written-out numbers.[94]

Ultimately, however, the advantages of Hindu-Arabic numerals
became too hard to ignore. Johannes Gutenberg's movable type, a
refinement of an age-old craft first developed in China, helped sta-
bilize the written forms of the new numbers. The invention of the
pencil in the eighteenth century and, in the nineteenth, the advent
of affordable, machine-made paper (another Western appropriation
of a Chinese invention) made a third act for the counting board an
impossibility.[95] The counting board was dead and the abacus's days
were numbered: the next step in the evolution of the pocket calculator was under way.

3 THE SLIDE RULE

THE EMPIRE STATE BUILDING, the Hoover Dam, and the Golden Gate Bridge owe their existence to it. So, too, the Spitfire, the steam train, Sputnik, and the atomic bomb. Sailors navigated with it; bombardiers bombed with it; and tax collectors calculated with it. It took us to the bottom of the ocean, and we took it to the surface of the moon. For generations of engineers, mathematicians, and scientists, the slide rule was *the* pocket calculator. And it would never have come about without the genius of a Scottish laird with a taste for the occult.

MOST WAYS OF looking at the world have some concept of modernity. There was a before and there is an after, and we recognize the after as being somehow closer to the current state of things. In the West, an art historian might point to the end of the nineteenth century, when artists began to make art and literature inspired by human experience rather than the whims of rich patrons. Modernism was the result.[1] For historians, the fall of Constantinople in 1453 is a handy bookmark for the start of what they call the Early Modern Era.[2] A scientist could cite the publication in 1543 of Nicolaus Copernicus's *De revolutionibus orbium coelestium* as kick-starting the Scientific Revolution.[3]

For mathematicians, the transition from before to after is not so much a point as an arc, described by a series of innovations in the late sixteenth and early seventeenth centuries. In Brittany, François Viète pioneered symbolic equations, where one can write, for example, $ax^2 + bx + c = 0$ without having to specify exact values for a, b or c.[*] Thomas Fincke, a Dane, and others began to explore trigonometric relationships that made it easier to solve geometric problems.[4] In Flanders, Simon Stevin promoted decimal fractions—that is, using the same notation after the decimal point as in front of it, rather than the traditional (and perplexing) system of switching to sexagesimal fractions.[5] Ludolph van Ceulen, a German emigrant to Holland, started with a regular pentadecagon (a fifteen-sided shape) and laboriously subdivided it thirty-seven times to come up with a thirty-five-digit approximation for π.[6] All these and more contributed to a sense that something was afoot in the world of mathematics.

And then there was John Napier, the Scot. Although Napier's contribution would lie firmly in the realm of capital-M mathematics, it could be applied to everyday arithmetic in a way that resonated like nothing before. If any of the little revolutions that punctuated the turn of the seventeenth century could be said to have brought mathematics into the modern era, it was John Napier's logarithms.

---

* Viète used Latin words, such as *A cubus*, rather than symbols such as A^3, but his heart was in the right place.

JOHN NAPIER WAS BORN IN Merchiston Castle, near Edinburgh, in 1550. His father, Archibald, a tender sixteen or under at the time of John's birth, was already the laird both of Merchiston and of a parcel of land in Stirlingshire, near Loch Lomond. Predictably, less is known about Napier's mother, Janet Bothwell, except that her brother Adam was installed as bishop of Orkney in 1559. It was Adam who urged the Napiers to send their firstborn abroad for his education, "for he can leyr na guid at hame, nor get na proffeit in this meist perullus worlde."[7]

And so, in the manner of upper-class families everywhere, in 1560 John was packed off to school in Europe. Where is not recorded (his uncle had suggested either France or Flanders), but Napier returned to Scotland in 1563 to enroll at the University of St. Andrews. Napier seems not to have finished his studies, but then a term or two at an elite university followed by an academic flameout is not an unfamiliar path for the children of the well-to-do.[8] As Napier grew into adulthood, the country of his birth was coming apart at the seams. James VI, the young king, was Protestant; his mother, Mary Queen of Scots, was Catholic, and forces loyal to mother and son still clashed on a regular basis. In 1572, John escaped a turbulent Edinburgh with his new wife, Elizabeth Stirling, and took up residence on the Napiers' northern estate.[9]

Ensconced in chilly Gartness, Napier did all the things a Scottish landowner might be expected to do.[10] He fathered twelve children, two with Elizabeth, who died in 1579, and ten more with his second wife, Agnes Chisholm. He managed the family estates and engaged in lawsuits about property. He took an interest in farming, patenting the use of salt as a fertilizer. And he wrote a book that accused the pope of being the Antichrist and put a date on the end of the world.[11]

That Napier could harbor such trenchant religious beliefs would not have been a surprise to those who knew him, since his father had been one of the first Scottish nobles to embrace the Reformation. Even so, the publication in 1593 of Napier's ultra-

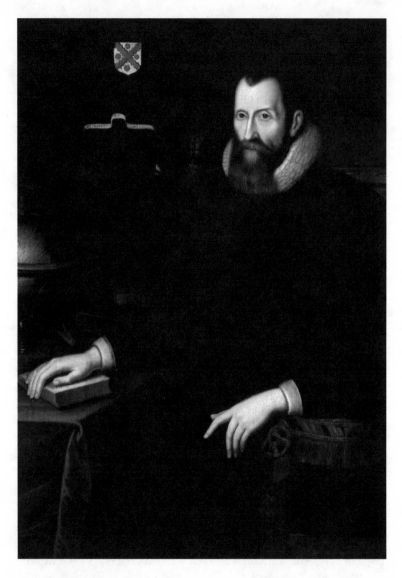

O John Napier in 1616, a year before his death, by an unknown artist. *Lord Napier of Merchiston (1616) by Unknown artist, oil on canvas, EU0001 © University of Edinburgh Art Collection.*

Protestant opus, *A Plaine Discovery of the Whole Revelation of Saint John*, was the first sign that Napier might be something other than a run-of-the-mill landlord with a passing interest in religion. In his book, Napier lambasted the papacy for thirteen hundred years of outrageous idolatry and sin and urged the king to rid his court of Catholics, atheists, and "newtrals." It was, on the whole, fairly standard anti-Catholic rhetoric, though it was interwoven with a startling prediction that the apocalypse would occur in either 1688 or 1700.[12] It was an unexpected hit: *A Plaine Discovery* was ultimately translated into French, German, and Dutch to better spread its message across Europe.[13]

In saving others' souls, Napier may have hoped to save his reputation. The year before *A Plaine Discovery* was published, Napier's father-in-law had been implicated in a conspiracy to bolster Catholic Spain's influence in Britain, and it may have been that Napier wanted to distance himself from his errant relation. The so-called Affair of the Spanish Blanks (named for blank letters signed by prominent Scottish Catholics) that had embroiled Napier's father-in-law was yet another symptom of Britain's undiminished religious strife:[14] In Scotland, open battles were being fought between Catholic and Protestant lairds.[15] England, under the Protestant queen Elizabeth, and Spain, under the Catholic Habsburgs, traded armadas like boxers trade punches.[16] Wherever one looked, and whichever side of the Reformation one chose, there was no room to stay "newtral." As such, Napier decided to involve himself in Britain's sectarian travails not only theologically, but practically too.

In 1595, James VI sent an ambassador to England to offer his assistance in Elizabeth I's ongoing war with Spain. Among the king's entreaties was a letter signed by one Jo. Neper of Marchistoun, detailing "secret inventions, profitable and necessary in these days for the defence of this island, and withstanding of strangers, enemies to God's truth, and religion." Jo. Neper of Marchistoun was, of course,

John Napier of Merchiston, who laid out for the king's perusal four engines of war with miraculous properties.[17]

Napier's first machine, a "round chariot of mettle [that] serveth to destroy the environed enemy by continuall charge and shott of harquebuse [a type of musket] through small hoalles," was a human-propelled proto-tank that would almost certainly have been stymied by its ponderous bulk.[18] The second device, a cannon of some sort, was apparently demonstrated at least once, though not in battle. In 1652, some years after Napier's death, Sir Thomas Urquhart wrote of the event:

> Of this it is said, that (upon a wager) [Napier] gave proof upon a large plaine in Scotland, to the destruction of a great many herds of cattel and flocks of sheep, whereof some were distant from other half a mile on all sides and some a whole mile.[19]

Pity the poor livestock who were (allegedly) blasted to bits to settle a bet. But that Urquhart, a notorious yarn-spinner who claimed descent from Bacchus and the Egyptian pharaohs, had prefaced his account with a precautionary "it is said" casts serious doubt on the reality of the second of Napier's machines.[20]

Finally, Napier gave an account of two mirrors designed to set fire to enemy ships, echoing Archimedes's supposed invention of the same.[21] One reflected and focused the sun's rays; the other made use of "any materiall fier or flame." Napier crowed that he could provide "proofe and [a] perfect demonstration, geometricall and alegebricall, of a burning mirrour" that focused the sun's rays into "one mathematicall point"*—which was all very well, but one wonders how

---

* With all of this sundry inventiveness, the reader might be forgiven for thinking of Napier as the quintessential Renaissance man. Napier himself would have

readily Napier could have turned his mathematical proof into a physical one in Scotland's famously dreich climate.[22]

All of this suggests that if Napier was not exactly a born warmonger, he was at least an inventive one. Likewise, his confidence in the mathematical correctness of his burning mirrors hinted at an affinity for numbers that far exceeded what might be expected of a provincial landowner. That affinity, in turn, came in handy for Napier's other hobby. For when he was not managing his estates, baiting Catholics, or designing war machines, John Napier was a committed student of the dark arts.

A DECADE BEFORE Napier's letter for the king, his father had been appointed master of the Scottish mint. There, the elder Napier oversaw the production of coins,[*] and it was there, most likely, that his son picked up a taste for alchemy.[23] After all, in an era of witch trials and religious fervor, goldsmithing and metallurgy must have seemed well-nigh coincident with the transmutation of base metals into gold.[24] Moreover, the Napiers had always been partial to a bit of light conjuring. John's father, Archibald, had been accused in court of invoking a demon in an esoteric tongue—Italian, as it turned out.[25] Adam Bothwell, John's uncle, somehow contrived to be both bishop of Orkney and a "sorcerer and execrable magician,"

---

balked at the label: after all, the Renaissance was a product of Catholic Italy, and Napier hated Catholics.

* Coincidentally, Johannes Gutenberg's father held a similar position in Mainz a century and a half earlier. It is interesting to compare what the two men did with their knowledge of metalworking: Gutenberg perfected movable type; Napier applied it to the practice of alchemy.

while an English cousin named Richard Napier acquired a bad reputation as an astrologer and alchemist.[26] Later in life, Napier himself would write detailed accounts of a series of meetings with a German alchemist named Daniel Müller, who gave the Scot the secret of the fabled Philosopher's Stone. Napier failed to manufacture this mythic substance, capable of transforming lead into gold, prolonging life, curing illness, and sundry other miracles, but then no one else succeeded either.[27]

In Napier's time, alchemy went hand in hand with astrology as part of what would today be called the natural sciences.[28] Nowadays, astrology is synonymous with TV mystics and tabloid horoscopes, but for Napier and his contemporaries, the study of celestial bodies and their influence on the natural world was as close to a scientific pursuit as had ever existed. Weather, tides, health, magnetism, war, fate—these and other phenomena were thought to be coupled to the motion of the heavens, which made astrology a worthy endeavor.[29] Luminaries such as Galileo Galilei, Johannes Kepler, Tycho Brahe, and Elizabeth I's court astronomer, John Dee, all carried out astrological predictions at one time or another. Even when astrology came under attack, as it increasingly did throughout Napier's lifetime, critics were more concerned that horoscopes might be correct and reveal unsavory truths, rather than with the validity of astrology itself.[30]

The counterintuitive thing about astrology is that it is hard. The mystical aspects may be nothing more than wishful thinking, but the observations and calculations on which they rely are embedded in cold, hard, messy reality. As such, to describe the motion of the spheres, ancient astrologers resorted to increasingly complex physical and mathematical concepts.[31] Around 350 BCE, for example, Aristotle postulated that the stars and planets must be fixed to some fifty-five concentric, interconnected spheres in order to give them their apparent motion.[32] A century and a half later, a philosopher named Apollonius added the idea of "epicycles," where each planet orbited a fixed point that itself orbited the Earth,* a fiction necessary to explain the planets' trajectories as seen from the ground.[33] Later still, in the second century BCE, a mathematician named Hipparchus constructed the first trigonometric tables to aid his astronomical calculations.[34] And so on down the centuries, all the way until 1594 or thereabouts, when John Napier had the first glimmers of what would become his own monumental contribution to astrological mathematics.[35]

NAPIER HAD LONG WORKED at the cutting edge of mathematics. *De Arte Logistica*, one of his earliest books on the subject, deftly handled negative numbers at a time when his peers barely acknowledged their existence. (Negative numbers had been discussed in earlier treatises written in Sanskrit, Chinese, and Arabic, but European mathematicians had been slow to take them up.) *De Arte* also introduced and then dismissed what are now called "imaginary numbers"—multiples of the square root of negative 1—that have since turned out to be indispensable to quantum mechanics, electrical engineering, and other mathematical disciplines.[36]

As such, when Napier learned of a new development in mathematics in the final years of the sixteenth century, he could not help but admire its cleverness. It was called *prosthaphaeresis*, and it allowed

---

* In 1687, seventy years after John Napier's death, Isaac Newton would put an end to more than a millennium of such outlandish theories with the observation that all physical bodies exhibit an attraction to one another in proportion to their mass. For all intents and purposes, the publication of *Principia Mathematica* was the alchemical act that transmuted astrology into astronomy.

its practitioners to multiply two numbers together without the need for actual multiplication. It looked like this:

$$\sin(a) \times \sin(b) = \frac{\cos(a - b) - \cos(a + b)}{2}$$

Although this seems complex at first glance, a mathematician of Napier's caliber would have appreciated its underlying simplicity. Here, sin and cos (sine and cosine, to give them their full names) are mathematical "functions," or equations, into which one can insert a value and receive a second value computed from the first. For example, the sine of 90 degrees is 1 and the sine of 180 degrees is 0. Sines and cosines relate to angular measurements, and although they are difficult to compute from scratch, by Napier's time it was possible to lay one's hands on books containing nothing but tables of precomputed sine and cosine values.[37]

To the astronomers, astrologers, and ships' navigators of Napier's time, who often had to carry out exactly this kind of angular multiplication, prosthaphaeresis was a godsend. This unheralded equation suddenly made it possible to convert vital but time-consuming multiplications into much simpler operations. Granted, one had to have tables of sines and cosines on hand, but then the sixteenth-century Venn diagram of people who needed to multiply the sines of two angles and people who owned books of trigonometric tables would have been very nearly a perfect circle.

As it turned out, prosthaphaeresis was even more useful than it at first appeared. Imagine you need to multiply any two numbers together. (Perhaps you are a merchant calculating interest rates rather than an astrologer computing when Mars will rise.) You can use prosthaphaeresis to do it. Broadly speaking, you must look up the two numbers to be multiplied in a table of sines to find the corresponding angles a and b, plug them into the right-hand side of the equation above, and work through to your result. Voilà: you have multiplied your two numbers without having carried out any multiplications at all.

Napier saw prosthaphaeresis, admired it, and mused that he could do better if only he gave it some thought. In the end, it took him twenty years of cogitation: he began working on the problem in the last decade of the sixteenth century and finished, finally, in 1614. The results of his work, a book entitled *Mirifici Logarithmorum Canonis Descriptio* (A Description of the Wonderful Law of Logarithms), was published that same year, and its author, now a venerable sixty-four years of age, would be dead three years later.[38] But no matter. With that one book, Napier's place in mathematical history was assured.

N THE WAY OF all of the best inventions, with hindsight Napier's surpassing breakthrough seems almost obvious. He started with a pair of simple, related concepts: arithmetic and geometric progressions. An arithmetic progression is a sequence of numbers where each number is computed by adding a fixed value to the one before it. For example, an arithmetic series that starts at 0 and has a "common difference" of 1 looks like this:

$$0, 1, 2, 3, 4, 5, 6, 7, 8, 9 \ldots$$

A geometric progression, on the other hand, is one where each number is computed by multiplying the preceding number by a fixed value. For example, a geometric progression that starts at 1 and has a "common ratio" of 2 looks like this:

$$1, 2, 4, 8, 16, 32, 64, 128, 256, 512 \ldots$$

Napier, along with other mathematicians of the time, already knew that it was possible to pair an arithmetic progression with a geometric progression such that certain multiplications could be converted into simple additions.[39] For instance, take the two progressions above and write them alongside each other, as shown in Table 1, opposite.

Next, to multiply any two members of the *geometric* progression, first look up their counterparts in the *arithmetic* progression. To multiply 4 and 8 from the geometric progression, for example, we see that they correspond to 2 and 3 respectively in the arithmetic progression. Then, we add the numbers from the arithmetic sequence and look up the corresponding number in the geometric sequence. In this case, 2 + 3 = 5 and the fifth entry in the geometric progression is 32. We have successfully multiplied 4 and 8 using only a single addition and a table of numbers. (Division is easy, too, but it uses subtraction rather than addition.)

It is a neat trick, but a limited one. What if we need to multiply two numbers that are not part of a given geometric progression? How can we multiply 25 and 89, for example, if they do not exist in our table? Really, it would be better if our geometric progression contained a lot more numbers—say, *all* of them—and that our arithmetic progression were filled in to match.

This was Napier's great innovation. He devised a mathematical function that took any number in a geometric progression, whether whole or fractional, and computed the corresponding arithmetic value.[40] At first he called these computed values "artificial numbers," but settled instead on the Greek *logarithm*, or "ratio number," named for the particular method by which he calculated them.[41]

To see how logarithms work, consider the first few values from Table 1. Using Napier's idea of logarithms, we can fill the gaps in the geometric series and then calculate corresponding values (which we call logarithms) for the arithmetic series. If we replace the geometric series with the whole numbers from 1 to 10, for example, and calcu-

TABLE 1

Arithmetic	0	1	2	3	4	5	6	7	8	9	...
Geometric	1	2	4	8	16	32	64	128	256	512	...

TABLE 2

Logarithm	0	1	1.585	2	2.322	2.585	2.807	3	3.17	3.322	...
Number	1	2	3	4	5	6	7	8	9	10	...

late the corresponding logarithms, we end up with the values shown in Table 2 on the previous page.*

The terminology has changed, but the principle is the same. To multiply any two numbers, first look up and add their corresponding logarithms, then find the number corresponding to the summed logarithm. For instance, to multiply 2 and 5, for example, we find their corresponding logarithms, as shown in Table 3, opposite.

Next, we add the logarithms together: 1 + 2.322 = 3.322. Finally, we look again at the table to find the resulting logarithmic value, as seen in Table 4.

Our summed logarithm, 3.322, corresponds to the number 10. We have determined that 2 × 5 = 10, all without having to carry out any multiplications at all. All that remains, to make our technique more and more useful, is to expand our table to encompass as many values as possible.

That was the theory, anyway, but Napier's articulation of it left something to be desired. As befitted an astrologer, with an astrologer's need to solve trigonometric problems, *Mirifici Logarithmorum Canonis Descriptio* waded straight into a dense, fifty-seven-page description of how logarithms could be used in two- and three-dimensional geometry. The tables of logarithms that filled the remaining ninety pages were obscured behind layers of trigonometric complications that made them essentially useless for day-to-day mathematics.[42] Yet even saddled with these drawbacks, *Descriptio*'s message rang out just as loudly as *A Plaine Discovery* had done in the previous century, if in slightly different circles of society.

---

* For the mathematical reader, these are base-2 logarithms, chosen to match our original table of arithmetic and geometric progressions.

TABLE 3

Logarithm	0	1	1.585	2	2.322	2.585	2.807	3	3.17	3.322	...
Number	1	2	3	4	5	6	7	8	9	10	...

TABLE 4

Logarithm	0	1	1.585	2	2.322	2.585	2.807	3	3.17	3.322	...
Number	1	2	3	4	5	6	7	8	9	10	...

O The first page of logarithms in Napier's *Mirifici Logarithmorum Canonis Descriptio*. The spiritual forerunner of the slide rule, if not quite so portable. *Napier, John, and Andro Hart. Mirifici Logarithmorum Canonis Descriptio, Ejusque Usus, in Utraque Trigonometria: Ut Etiam in Omni Logistica Mathematica: Amplissimi, Facillimi, & Expeditissimi Explicatio. Edinburgi: Ex officina Andreae Hart, 1614. Pdf. https://www.loc.gov/item/04005707/.*

LOGARITHMANIA* BEGAN IN England with Henry Briggs, a renowned professor of geometry in London, who came to Merchiston the year after Napier published *Descriptio* to pay hom-

---

* Believe it or not, I am far from the first person to make this pun.

age to the Scottish laird.[43] (Napier knew Briggs by reputation, and the almost-certainly-apocryphal story goes that, on meeting, the two scientists gazed admiringly at each other for a quarter of an hour before they thought to break the silence.[44]) At Napier's suggestion, Briggs calculated logarithms for the numbers 1 to 1,000, making them more usable in everyday multiplication, and published them in 1617 in a sixteen-page pamphlet.[45] In 1624, the famed German astronomer Johannes Kepler published his own set of logarithms; a year later, the first French translation of Briggs's pamphlet was printed in Paris; and by 1628, a Dutchman named Adrian Vlacq had published a table of logarithms for the numbers 1 to 101,000.[46] Barely two decades after Napier first announced his discovery, logarithms were in use across the world.[47] Gunnery, navigation, astronomy, surveying, tax collection, banking—if a profession needed to multiply or divide, it needed logarithms.

But not all was rosy in the new logarithmic world. Even with Henry Briggs's improvements and the compendious tables that followed, one still had to carry around a book in order to make use of them. The printed book as an aid to calculation was not a new idea—the first such example, containing tables of compound interest rates, pre-dated Napier's efforts by three decades, not to mention the Sumerians' mathematical tables of millennia before—but multiplication and division were too useful to be confined to a desk. Logarithms were crying out for a more portable home.[48]

Enter Edmund Gunter. A clergyman and friend of Henry Briggs, Gunter had a knack for designing physical tools that embodied complex mathematical ideas. Sundials were his first love, but he also improved the venerable astrolabe, a device for measuring inclination; pioneered the use of metal chains to survey distances; and measured ships' speeds with a knotted line.[49] As a case in point, in 1620, when Gunter arrived at a job interview at Merton College in Oxford, he took out navigational tools used at sea and began demonstrating how to use them.[50] The warden of the college, Sir Henry Savile, was unimpressed:

Doe you call this reading of Geometrie? This is shewing of tricks, man![51]

Gunter did not get the job. Instead, he and Briggs swapped institutions: Gunter went to Gresham College, where Henry Briggs had worked since 1597, and Savile gave Briggs the nod at Oxford.[52]

At Gresham College, a tiny institution dedicated to bringing science to the people of London, Gunter's main responsibility was to deliver twice-weekly public lectures on astronomy.[53] It left him plenty of time to indulge his passion for scientific instruments, and so, in 1624, Gunter published a landmark book entitled *The Description and Use of the Sector, the Crosse-Staffe and other such Instruments*.[54] Written in vernacular English rather than scholarly Latin, it did more to bring logarithms to the masses than any work before or since.[55]

The instruments of the book's title were already well known to astronomers and navigators. The cross-staff was a relatively simple device for measuring angles, such as the elevation of stars above the horizon, but the sector was quite different. At least three versions had been independently invented during the second half of the sixteenth century, one of them by none other than Galileo Galilei, all of which resembled a folding ruler inscribed with numeric scales. All sectors operated on the same principle: that the sides of any two triangles with the same angles will always be in perfect proportion.[56] It was this property that allowed a sector to be set up to multiply and divide repeatedly by a factor chosen by its user—but only after a laborious process of guesswork, measurement, and at least two other multiplications carried out on paper or in the user's head. It was so fiddly that many sectors, acquired in fits of enthusiasm that waned as their owners tried to actually use them, have survived to this day in mint condition.[57]

Gunter knew the sector could be improved, just as Napier knew prosthaphaeresis was only a stepping-stone to something better. Having dealt with the sector in the first part of his book, in the sec-

O Galileo's "geometrical and military compass," more commonly identified as a sector. *Mazzoleni, Marc'Antonio. "Galileo's Geometrical and Military Compass (DW0950)." Collection of Historical Scientific Instruments. Cambridge, MA: Harvard University, 1603. http://waywiser.fas.harvard.edu/objects/3608/galileos -geometrical-and-military-compass. Image courtesy of Collection of Historical Scientific Instruments, Harvard University.*

ond Gunter proposed a radical new instrument that could multiply and divide with preternatural ease. Taking inspiration but little else from the sector, Gunter made the case for what he called a "line of numbers"—a physical scale, around two feet long, on which numbers were located according to their logarithms. That is, rather than placing the numbers 1 to 100 equally along the scale like a ruler, Gunter looked up their logarithms and used them to decide where to place the numbers, so that the divisions on his scale started off widely spaced on the left and grew more crowded to the right.[58]

To use Gunter's scale to multiply any two numbers was simple. First, measure the distance from 1 to the first number. Then, move the measurement so that it starts at the second number to be mul-

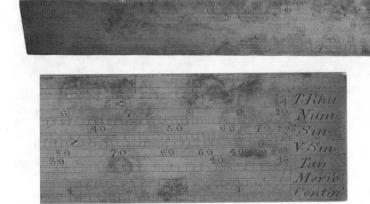

O A brass Gunter's scale of the early nineteenth century. In common with many sectors and similar instruments, this rule has multiple scales for computing different values. Edmund Gunter's great innovation was the logarithmic scale labeled "NUM." *Berge, Matthew. "Brass Gunter's Rule (DW0564)." Collection of Historical Scientific Instruments. Cambridge, MA: Harvard University, 1810. http:// waywiser.fas.harvard.edu/objects/12404/brass-gunters-rule. Image courtesy of Collection of Historical Scientific Instruments, Harvard University.*

tiplied, thus adding the logarithms of the two numbers. The result of the multiplication lies at the end of the measurement.[59] It was a rough-and-ready device, producing answers accurate only to a few significant figures, but that aside, Gunter's line of numbers was so much better than the sector that it supplanted it almost immediately.*

"Gunter rules" became staple tools for ships' navigators for more

---

* Gunter's scale could multiply only numbers within its inscribed range. A savvy mathematician, however, could shift decimal points around to get around that limitation. A multiplication by 314, for example, could be recast by shifting the decimal point one place to the left to yield a multiplication by 31.4, or two places to give 3.14. Results were then scaled up by a factor of 10 or 100 as appropriate, shifting the decimal point to the right to get the correct answer.

than three hundred years. Inscribed on a simple piece of boxwood, with no moving parts to seize or rattle at sea, a logarithmic scale could help plot courses across Gerardus Mercator's famous charts; it simplified astronomical calculations when navigating by the light of the stars; and it speeded up tricky calculations when in dangerous coastal waters.[60] For the landlubbers, however, there was one more logarithmic innovation yet in store.

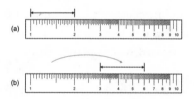

O   How to use a Gunter scale: Use a pair of dividers to (a) measure from 1 to the first number, then (b) move the dividers so that the left-hand point lies at the second number. The right-hand point shows the final product. In this case, 2 × 3 = 6.

L IKE EDMUND GUNTER, William Oughtred was a preacher with a flair for mathematics. He moved in the same circles, too, visiting Henry Briggs at Gresham College and meeting Gunter on at least one of those visits.[61] Unlike Gunter, however, Oughtred was wary of using physical devices to solve mathematical problems, writing that such baubles turned mathematicians into "doers of tricks, and as it were Iuglers [jugglers]." He bewailed the "superficiall scumme and froth of Instrumentall tricks and practices."[62] Ironic, then, that Oughtred would invent the most important calculating device of the next three centuries.

Oughtred was at the heart of the mathematical and scientific revolution sweeping across Europe. His parishioners in the village of Albury, near London, complained of dull sermons informed more by math than by ministry; little wonder, given that Oughtred was distracted by his role as one of the key figures in the exchange of scientific letters across the continent. Aspiring mathematicians were drawn to him like moths to a flame.[63]

In 1630, so the story goes, six years after Edmund Gunter's seminal *Description and Use of the Sector, the Crosse-Staffe and other such Instruments*, one of Oughtred's students mentioned how time-consuming and inaccurate it was to use Gunter's line of numbers. That student, William Forster, had commissioned a scale fully six feet long that he used with a giant pair of dividers in order to measure off more accurate results. Unexpectedly, Oughtred showed Forster how to slide two Gunter rules against each other to do away with dividers altogether, and just like that, the principle of the slide

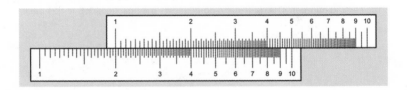

O   How to use a slide rule: Slide the top rule so that the 1, or "index," lies above the first number to be multiplied on the bottom rule. Find the second number to be multiplied on the top rule and read the product immediately below it on the second rule. In this case, the index on the top rule is aligned with the number 2 on the bottom rule. By finding 3 on the top rule, for example, and looking down to the bottom rule, we see that 2 × 3 = 6.

rule was born.[64] (Oughtred's sliding rules are as difficult to describe as they are easy to use: see the figure opposite for a pictorial guide on how to use them.)

Next, Oughtred showed Forster an instrument he called the "circles of proportion," comprising two overlaid disks with logarithmic scales engraved around their perimeters. It worked on the same principle as his sliding rules: rather than use dividers to measure one distance and add it to another, the two logarithmic scales were rotated against each other to do the same. And because one could wrap a longer scale around a circle than along a ruler of the same width, more accurate readings could be taken.[65] As with his sliding rules,

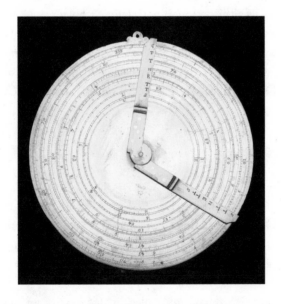

O A mid-seventeenth-century circular slide rule, made in London, England. The scale labeled "N," for "NUM," is logarithmic. *Collection of Historical Scientific Instruments. "Circular Slide Rule with Oughtred-Type Sundial Reverse (DW0959)." Cambridge, MA: Harvard University. Accessed October 7, 2020. http://waywiser.fas. harvard.edu/objects/2818/circular-slide-rule-with-oughtredtype-sundial-reverse. Image courtesy of Collection of Historical Scientific Instruments, Harvard University.*

Oughtred's circles of proportion were simple, and brilliant, and decidedly out of character for the pen-and-ink purist.

Oughtred had not publicized his inventions, he told Forster, because he did not care for such toys.[66] All that changed when another of his students, Richard Delamain, published a description of Oughtred's instruments and claimed them for his own. Oughtred fired back with a scathing response that included his deathless barb about "doers of tricks and [. . .] Iuglers," but he was firmly on the wrong side of history. For many would-be mathematicians, mastering the "superficiall scumme and froth of Instrumentall tricks" was easier than learning the principles that underpinned such instruments, and in any case, whipping out a sector or Gunter's scale was a surefire way to impress a wealthy patron, if not a fellow scholar. As a case in point, Delamain used his knowledge of mathematical instruments to wheedle his way into the good graces of Charles I, king of England and Scotland, while Oughtred singularly failed to do the same.[67] Whether William Oughtred liked it or not, his sliding rules were here to stay.

WITH ITS MAIN INGREDIENTS now in place—John Napier's logarithms, Edmund Gunter's line of numbers, and William Oughtred's sliding rules—the slide rule took shape. And in contrast to the theoreticians who had laid down its foundations, it would be its users who would transform the slide rule into a practical tool for all.

The first real advance took place in 1657, when an English surveyor named Seth Partridge refined Oughtred's pair of sliding rules into a single instrument comprising two fixed "stocks" and a movable slide that ran between them.[68] Next came the idea of the specialized slide rule: In 1677, one Henry Coggeshall described a folding wooden ruler with a built-in slide rule that helped carpenters and stonemasons to measure and calculate areas and volumes of timber and stone. It was innovation of a very modest sort, but it was enough to bring the slide rule into the public eye.[69]

More intriguing still was Thomas Everard's 1683 "gauging rule." Gauging was the art of determining the "wet" and "dry" volumes of a barrel so that its contents could be accurately taxed, and it was not for the mathematically timid.[70] As such, Everard's adapted slide rule was bursting with novel features: It was double-sided, with dedicated slides and multiple logarithmic scales on both front and back faces. Those scales were peppered with brass pins marking constants such as pi and $\sqrt{2}/2$, and also more practical units, such as the number of cubic inches in a malt bushel or a gallon of ale (2150.42 and 282 respectively, but you knew that).[71] A reversed scale, counting down from 10 to 1, made it easy to calculate reciprocals (that is, $1/x$), while scales running from 1 to 100 and 1 to 1,000 could be used to compute squares, square roots, cubes, and cube roots.[72] Everard's extra scales were not a new idea— many sectors and Gunter rules already boasted multiple scales—and, strictly speaking, none of them relied on logarithms, but nevertheless they made his slide rule considerably more versatile.

There was one more innovation to come, after which the form of the slide rule was essentially fixed. The so-called cursor, a sliding vertical bar that made it easier to read slide rules with multiple scales, was independently invented at least four times before it finally stuck. First to come up with the idea was Isaac Newton, shortly after logarithmic scales became fashionable, but it was a young French artillery officer named Amédée Mannheim whose 1850 design, bearing a metal cursor and four scales labeled A to D, became the prototype for all that followed.[73]

The end result of all this tinkering and tweaking—this elaborate ruler, this clever stick—was a portable mathematics powerhouse. Five minutes' instruction with a slide rule turned it from a maze of

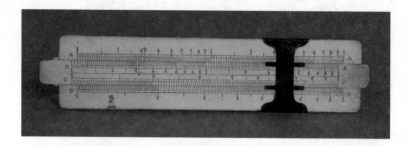

O An early-twentieth-century slide rule with a Mannheim-style cursor, allowing multiple scales to be cross-referenced with greater accuracy. *"Gebr. Wichmann Mannheim Simplex Slide Rule." National Museum of American History. Washington, DC: Smithsonian Institution, n.d. https://americanhistory.si.edu/collections/ search/object/nmah_690256. Image courtesy of Division of Medicine and Science, National Museum of American History, Smithsonian Institution.*

confusing scales into a magic wand that calculated squares, cubes, and roots as easily as it multiplied and divided. There was no juggling of ratios or fussing with dividers, just a flick of the slide here and a nudge of the cursor there. It must have been a revelation. And it showed: Coggeshall's carpenter's rule was still popular two centuries after its invention; excise officers continued to use Everard's gauging rule into the twentieth century; and Mannheim-style slide rules were the first to entice reluctant American buyers in the late nineteenth century.[74]

A S THE SLIDE RULE GREW in popularity, there followed a steady stream of variants for any discipline that bore even a hint of empiricism. Rabone & Son's 1878 "cattle gauge" was a circular slide rule with a built-in measuring tape, the better to estimate a cow's weight from its length and girth.[75] The fin-de-siècle "actinograph" was used by photographers to calculate light levels at different times of day, dates, and latitudes.[76] From early in the twentieth century, the Met Office, Britain's weather forecasting service, used specialist slide rules to derive wind velocities from the observed flight paths of weather balloons.[77] Gilson's circular "Commercial Calculator" computed interest on loans.[78] The Fisher Governor Company of Marshalltown, Iowa, commissioned slide rules that helped its engineers determine the size of valves required to manage flows of gas or liquids.[79] Even Disney could not resist, churning out a slide rule–inspired adding device emblazoned with Mickey Mouse's grinning face.[80]

There was an especially rich vein of slide rules in aviation and aerial warfare. In the 1930s, a circular slide rule called the E6B was created to help pilots calculate fuel burn, estimate ground speed, convert between different units, and more. The E6B appeared in *Star Trek* more than once, a retro-futuristic anachronism in Mr. Spock's scientific toolkit, and it is still in use for training pilots a century after its invention.[81] (In 1952, Swiss watch makers Breitling released a watch with an E6B-inspired scale inscribed around its movable bezel.)[82] In World War II, flight engineers and navigators used slide rules with interchangeable celluloid faces so that different problems could be solved using appropriate scales.[83] Bombardiers carried specialized slide rules, too, as backups to their intricate mechanical bomb sights.[84]

As if to underscore its utter, inescapable ubiquity, the slide rule was complicit in the dreadful events over Hiroshima, Japan, on August 6, 1945. At least two slide rules flew aboard the *Enola Gay* that day as it dropped "Little Boy," the first atomic bomb—a bomb that had been created by physicists, mathematicians, and engineers who wielded slide rules as essential tools of their respective trades.[85] In the years afterward, a macabre strain of slide rules calculated the explosive power, expected damage, and radioactive fallout that would be dealt by Little Boy's descendants, if ever they were used in anger.[86] John Napier's human-powered tank and ship-burning mirrors may have come to naught, but his invention of logarithms was a necessary precondition for the most destructive weapon humanity has ever devised, and much else besides: from taxing beer to splitting the atom, the slide rule ruled the greater part of the modern era.

BY THE MIDDLE OF the 1960s, more than a million slide rules were being sold each year in the United States alone. The Pickett Corporation of Chicago was one of the leading U.S. manufacturers, and it was in the form of the diminutive Pickett N600-ES that the slide rule would reach its peak.[87] Physically speaking, the N600-ES was not without quirks, though only the most devout of slide rule enthusiasts would lose much sleep over them. It was only five inches in length, for example, whereas most standard rules were twice as long. It was a relaxing yellow color, too (hence its "ES" suffix, for "Eye Saver"), and it was made of aluminum rather than the more usual celluloid and wood.[88] Those features aside, it was not a slide rule to challenge the status quo. Amédée Mannheim would have been able to pick up and start calculating with an N600-ES with no trouble at all.

But that is not why the N600-ES is interesting. It is interesting because NASA decided that it was worth expending 2,343 pounds of rocket fuel to put it on the moon. Somewhere between five and ten N600-ESs were fired into space, with one or two examples flying on each of the first five Apollo missions as backups for the electronic computers that guided and controlled the spacecraft.[89] "No modifications were needed for use in space," the Smithsonian notes drily of the space-flown N600 in its collection, because, let us not forget, the twenty-two scale, double-sided, trigonometric Pickett N600-ES was scarcely more complex than an elementary school ruler.[90] It is not known whether Neil Armstrong or Buzz Aldrin ever had to break out their slide rules while aboard *Apollo 11* (Pickett Corporation, who crowed mightily about their lunar rules, would certainly have let their customers know about it), but that did not stop Aldrin's rule from fetching $77,675 at auction in 2007.[91]

From such dizzying heights, the slide rule had plenty of time to watch the ground rush up to meet it.

No one would ever claim that the slide rule was perfect. The foot-long models used by most engineers could produce answers accurate only to a few significant digits—2.43, say, or 0.243, or 0.0000243, depending on where the user put the decimal point in their calculations. (Longer rules, along with circular and cylindrical versions, were more precise but less speedy.) What was worse, because slide rules were of limited use in solving equations involving squares, cubes, and so on, there was a temptation—even a need—to simplify, simplify, simplify until problems could be solved with nothing more complex than addition, subtraction, multiplication, and division. As a result, bridges were built stronger, and more expensively, than they

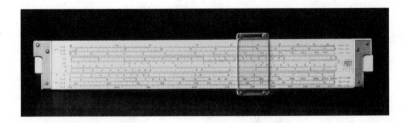

○ A slide rule owned by Dr. Sally K. Ride, the first American woman in space. Ride first flew aboard the Space Shuttle *Challenger* in 1983; whether her slide rule went with her is not recorded. *National Air and Space Museum. "Slide Rule, Mathematical, Sally Ride." Smithsonian Institution. Accessed September 15, 2020. https://www.si.edu/object/slide-rule-mathematical-sally-ride:nasm_A20140276000. CC0 image courtesy of National Air and Space Museum, Smithsonian Museum. Gift of Tam O'Shaughnessy.*

had to be. Planes were heavier and slower. Cars burned more gas. Oceans of nuance and complexity were erased.

But the final nail in the slide rule's complimentary pleather belt holster was not that it was inaccurate, or that it could only address a limited set of mathematical functions. The real problem was that it asked too much of its user: it put them closer to the numbers, yes, but it also asked them to estimate and fudge and squint and recompute. As such, when a viable alternative arrived, the end was swift.[92] Just three years after Neil Armstrong and Buzz Aldrin carried their Pickett slide rules to the moon, the astronauts who flew to the Skylab space station took HP-35 electronic calculators instead. In 1976, Pickett stopped making slide rules; in 1980 their remaining inventory was sold for scrap.[93] William Oughtred's clever stick was no match for the scientific calculator.

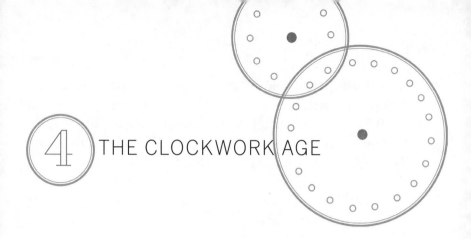

4 THE CLOCKWORK AGE

I
N THE SEVENTEENTH CENTURY, Europe's moneyed classes developed a mania for machines. They synchronized their pocket watches to pendulum clocks; they marveled at clockwork toys such as genuflecting monks and arrow-firing goddesses; and they listened to melodies played by automated organs and harpsichords.[1] By 1664, the philosopher René Descartes could write that animals were merely machines that behaved as "every bit as naturally as the movement of a clock, or other automaton," safe in the knowledge that his readers could nod their heads in understanding, if not in agreement.[2]

This was all part of a lineage of mechanical contraptions stretching back to ancient Greece and Rome, and whose branches encompassed Europe, Africa, and Asia.[3] Among the ancestors of the gold and brass playthings of Europe's elite were Archimedes's water clocks; Hero's spherical *aeolipile*, or rotating steam engine; and the sundry drinks dispensers, clockwork boats, and musical instruments described in the thirteenth century *Book of Ingenious Mechanical Devices*, written in what would become modern Turkey.[4] Closer to home in Europe, a parade of cathedral clocks, fountains, waterworks, and other mechanical diversions reached its peak in the sixteenth century with the advent of "androids"—clockwork simulacra of humans or animals.[5]

With the algorists in the ascendance, the counting board on its last legs, and mechanical marvels multiplying all the while, the time was ripe for a new, technological approach to calculation.

THE GERMAN TOWN of Tübingen lies at the western point of an equilateral triangle anchored by Augsburg and Nuremberg, two cities famed for their clockwork masterpieces,* and it was here that the pocket calculator would take its next evolutionary step.[6] In 1609, Tübingen's university awarded a degree in languages and theology to a student, Wilhelm Schickard, who hailed from a family of artisans. Schickard was something of a prodigy, having previously won a scholarship to the prestigious Tübinger Stift,† a seminary endowed by the Duke of Württemberg, and he would go on to acquire a master's degree two years later.[7] Schickard loafed around the university for another two years after that, until, in 1613, he found a position in a local church.[8]

At the university, Schickard had studied under Michael Mästlin, an eminent mathematician and astronomer, and it was during Schickard's short career as a Lutheran minister that he made the acquaintance of another of Mästlin's students. Two decades older

---

* For the clockmakers' guilds of the time, "masterpiece" had a deceptive meaning: it referred to a clock submitted by a prospective guild member to gain admittance. Masterpieces were typically conservative rather than innovative so that there were no untested movements to go wrong. In this sense, a clockmaker's best work was most often *not* found in a masterpiece.

† Students of the Tübinger Stift are called "Stiftlers." This should not be funny, but it is.

than Schickard, Johannes Kepler had already spent a stint as the official mathematician to the Holy Roman Emperor and had published a number of groundbreaking texts on the mechanics of the solar system. By 1615, however, with Kepler's imperial patron having been deposed some years earlier, the astronomer's star was on the wane. Not only that, but Kepler had been forced to return to Tübingen to defend his mother, accused of witchcraft, in the courts.[9] For Kepler, it was a trying time indeed, but for Schickard, it was a chance to meet one of the leading scientific minds of his era.

By all accounts, the two hit it off.[10] Kepler saw a kindred spirit in Schickard and encouraged the younger man, already a creditable painter, engraver, and mechanic, to develop his abilities as an astronomer and mathematician. He showed Schickard tables of John Napier's logarithms, then still a novelty, and asked Schickard to engrave some of the copper plates for a book on planetary motion.[11] And it was in the course of the many letters exchanged between the two that Schickard revealed a side project inspired by an astronomer's need to crunch large quantities of large numbers. On September 20, 1623, Schickard described to Kepler what he called an *arithmeticum organum*, or arithmetical instrument:

> What you have done in a logistical way (i.e., by calculation), I have just tried to do by mechanics. I have constructed a machine consisting of eleven complete and six incomplete (actually "mutilated") sprocket wheels which can calculate. You would burst out laughing if you were present to see how it carries by itself from one column of tens to the next or borrows from them during subtraction.[12]

Freeze frame, record scratch, et cetera. This is the first written description of a mechanical calculating machine—a device that encodes the rules of arithmetic within its own mechanisms rather than delegating them to its human user. Schickard sent Kepler another letter in February the following year, presumably at the latter's request, in which he described the machine's design and operation in much greater detail. Kepler would have to be content with Schickard's words alone: a machine intended for Kepler was lost in a fire, and Schickard's own prototype has never been found.[13]

And that might have been that, but for the fact that drawings of Schickard's *Rechenuhr*, or "calculating clock," as it is usually called, were discovered by chance in the library of an observatory near Leningrad (now St. Petersburg) in Russia. There, in 1957, while researching a biography of Johannes Kepler, a historian named Franz Hammer discovered a slip of paper that carried Schickard's origi-

O A 1957 replica of Wilhelm Schickard's *Rechenuhr*, or "calculating clock." The manually operated rods on top help the user deconstruct difficult multiplications and the clockwork section at bottom adds up individual components of those multiplications. *Freytag-Löringhoff, Baron von. "'Rechenuhr' von Wilhelm Schickard." Heinz Nixdorf MuseumsForum, 1957. https://nat.museum-digital.de/object/69093?navlang=de. © Sergei Magel, Heinz Nixdorf MuseumsForum*

nal sketches.[14] Hammer announced his discovery in, of all places, a magazine dedicated to office equipment, and, in doing so, inspired a professor of philosophy in modern-day Tübingen to build a replica of Schickard's machine.[15]

The completed *Rechenuhr* had two independent parts. At the top was a set of six numbered wooden rods, called "Napier's bones," with which a practiced user could break a large multiplication into partial products. They were named for John Napier, who had published a description of them in the year of his death, but he had done so only to claim priority for an invention that was already becoming popular as a way to simplify large multiplications; the actual inventor, some centuries earlier, had been the famed Persian mathematician al-Khwārizmī. Ironically, Napier's bones would be so thoroughly eclipsed by the slide rule that their inventor's name is more often associated with a calculating device he did not invent than one he did.[16]

As used in Schickard's machine, the idea was that the rods could be used to find the partial products for a multiplication, which could then be added up using the lower part of the device. This was Schickard's real innovation: an honest-to-God mechanical adding machine.

Operating the *Rechenuhr*'s adding machine was simple. A set of digits visible in six windows recorded a running total, which was incremented by units, tens, hundreds, and so on by turning one of six associated dials. (Subtractions were enacted by turning the dials in the opposite direction.) Whenever a digit ticked over from 9 to 0, one of the "mutilated sprockets" that Schickard had described to Kepler, each one bearing only a single tooth, caused the adjacent digit to be incremented by 1. If that also ticked over, another 1 was carried to the next digit, and so on. In the worst case, adding 1 to a running total of 999,999 would cause all dials to carry, so that the single-toothed gear on the first dial strained to turn over the five remaining dials all at once. Schickard seems to have been aware of the problem, which may be why his machine had only six digits. His solution was to have a bell ring every time the most significant digit "overflowed," so that anyone carrying out a larger multiplication—say, an astronomer such as Kepler himself—could keep track of the excess not shown on the device's dials.[17]

Little came of Schickard's invention during his lifetime. The fire that claimed Kepler's intended copy of the *Rechenuhr* seems to have snuffed out any ambition Schickard may have had to develop it further, and the man himself perished in 1635 along with his family in a plague brought by invading Catholic troops.[18] It speaks volumes that Kepler's name has been applied to such things as a spaceborne telescope, a lunar crater, a Martian crater, a pair of stars, a supernova, an opera, and many other things besides, while Wilhelm Schickard makes do with only a vocational school and a university department in his home town of Tübingen.

THE LIFE AND WORKS of Blaise Pascal, the French polymath after whom are named Pascal's triangle, Pascal's theorem, Pascal's law, Pascal's wager, the Pascal programming language, and a unit of pressure called, yes, the pascal, are even more celebrated than those of Johannes Kepler.[19] For many years after his death, Pascal was also heralded as the inventor of the mechanical calculator, and, had Wilhelm Schickard's *Rechenuhr* not been rediscovered in 1957, there is every chance that he would still hold that title. Even dethroned, however, Pascal's machine, the "Pascaline," is as intriguing as the *Rechenuhr*. The man himself, even more so.

Born in 1623, Pascal, like Schickard, grew up in a Europe groping its way toward new kinds of philosophical and scientific thought. Unusually for the time, Blaise was home-schooled by his father, a keen mathematician—but curiously, his father forbade him from tackling the subject until he was fifteen years of age. When, at twelve years old, Blaise deduced that the angles in a triangle always add up to 180 degrees, his father relented.[20] Two years later, Étienne Pascal started bringing his precocious son to conclaves organized by one Marin Mersenne, a Catholic monk with a talent for math, in which the great and the good of Parisian science presented their ideas.

Like William Oughtred in England, Mersenne was at the heart of a network of natural philosophers that stretched across France. He was no mean mathematician, either, concerned as he was with the formula $n = 2^p - 1$, where, if n is a prime number (that is, a number divisible only by itself and by 1) then p must also be a prime number.[21] The largest known prime number ($2^{82,589,933} - 1$) fits the definition of n, making it a so-called Mersenne prime, and Mersenne's formula continues to intrigue researchers working in encryption techniques.[22]

In June 1639, to a group of eminent mathematicians and astronomers, Blaise presented the first advance in geometric math since the time of Euclid himself. Blaise's theory, then called *Hexagrammum Mysticum* and now known as Pascal's theorem, relates to how lines intersect on a slice taken through a cone. It is not the most accessible or practical of concepts, but it was clear that even at only sixteen Pascal was a force to be reckoned with.

The same year, Pascal's father was appointed tax collector of Rouen, a town a hundred kilometers from Paris.[23] Blaise continued to broaden his scientific horizons in his new home, and in 1648, he orchestrated one of the most celebrated experiments of the ongoing Scientific Revolution. Pascal dispatched his brother-in-law to the top of a dormant volcano called Puy de Dôme with a mercury barometer and instructed him to record any changes in the indicated pressure. A monk at the foot of the mountain was given similar instructions, and, when Pascal later compared both sets of readings, it confirmed his hypothesis that air pressure dropped with altitude—and that eventually, one might expect to encounter a void in which there was no air whatsoever.[24] Pascal explained his theory to René Descartes, another of Mersenne's circle, who had visited Rouen the previous year. Unimpressed and unconvinced, Descartes wrote of the encounter that Pascal "has too much vacuum in his head."[25]

I T WAS IN ROUEN that Pascal embarked on perhaps the least successful of all his intellectual ventures. But for Pascal, failure was relative. His calculating machine would stand unchallenged for three hundred years as the first of its kind.

As early as 1640, only a year after his father had taken up his new job, Blaise started to think about how to mechanize arithmetical operations in the hope of saving himself from the lengthy, tedious tax calculations demanded of him by his father.[26] He envisaged a machine that, he wrote later, would embody the rules of arithmetic *sans plume et sans jeton*—"without pen and without counting token."[27]

Bringing the machine to fruition tested Pascal almost beyond his abilities, and by 1642, he was on the cusp of giving up. Somehow, though, he found a guardian angel: Pierre Séguier, the head of France's judiciary, learned of Pascal's endeavors and encouraged him to persevere. Three years and fifty prototypes later, "some made up of straight rods [. . .], others of curves, others with chains, some moving in a straight line, others circularly, some in cones, others in cylinders, and others all different from these, either for the material, or for the shape, or for the movement," it was Séguier to whom Pascal dedicated his completed machine.[28]

The workings of the "Pascaline," as it came to be known, were never adequately documented by its inventor. Unlike its predecessor, the *Rechenuhr*, however, there is no question of whether Pascal's machine was ever actually built. At least ten Pascalines survive to this day, and although individual models vary in the details, they all work the same way.[29]

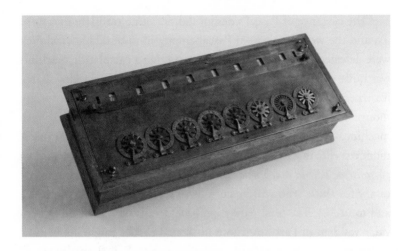

O   A modern replica of the Pascaline gifted to Queen Christina of Sweden in 1652.
Each dial enters a digit to be added, and each window records the current total
for that digit. This particular Pascaline is intended for monetary calculations: the
first six dials accept decimal inputs but the last two, labeled "Sols" and "Deniers"
respectively, are numbered from 0 to 19 and 0 to 12 so that they correspond
to France's pre-decimalization currency. *Nilsson, Robert S. "Räknemaskin."
Stockholm: Tekniska Museet, 1942. https://digitaltmuseum.se/021026304603/
raknemaskin. CC0 image courtesy of Nina Cronestrand / Tekniska Museet.*

Like Schickard's device, the Pascaline was essentially an adding
machine. It was operated via a number of dials, one per decimal position, and it maintained a running total visible through small windows
on the face of the machine. Inside, Pascal had used miniaturized
"lantern" gears, of a type more commonly found in large clocks, that
could handle high torques better than the *Rechenuhr*'s simple spur
gears. To these Pascal added a gravity-driven carry mechanism so
that no single dial had to turn over all the digits at once. Instead,
each dial had a weight that was raised by degrees as the user turned
it, and which, when the count ticked over from 9 to 0, dropped down
and incremented the next dial. It was an elegant workaround for the
"cascading carry" problem that had irked Schickard.

The elegance of Pascal's design did not quite translate into reality, and the workmen whom Pascal employed to build his machines were defeated by their intricacy. Constructing a calculating machine was quite different from building a house or a forging a plowshare, it turned out, and Pascal found himself learning the basics of iron-mongery and clockmaking in his quest to build a functional, robust machine.[30] Finally, in 1645, Pascal revealed his device to France's moneyed elite in a pamphlet entitled *Advis nécessaire à ceux qui auront curiosité de voir la machine arithmétique, et de s'en server* (Advice necessary for those who will be curious to see the arithmetic machine, and to use it). The final paragraph gave details of a Sieur de Roberval, a resident of Paris, to whom interested parties might apply for a demonstration. Pascal waited for the orders to roll in. They did not.[31]

WHAT, THEN, was wrong with the Pascaline? Inasmuch as it could add up whole numbers, it was a perfectly serviceable machine. Unfortunately, that was all there was to it. Subtraction was possible, but it occasioned some mental gymnastics and, thanks to the Pascaline's sophisticated carry mechanism, needed special handling of the machine itself. Multiplication meant adding and re-adding the number to be multiplied: an accomplished mathematician would likely have turned instead to pen and ink, a table of logarithms, or even, perhaps, to one of the "sliding rules" that held England's intelligentsia so rapt.[32]

Its creator may not have appreciated the comparison, but the truth was that the Pascaline inhabited the same niche as the clockwork toys that amused the rich. It was less ornate than most table clocks, certainly, and considerably less entertaining than the clockwork huntsman, currently held by the British Museum, which promenaded down the dining table carrying a slug of wine to be drunk by the guest in front of whom it stopped. Nevertheless, the Pascaline held the same kind of fascination, and carried the same lofty price tag, as those gaudier trinkets.[33]

The wording of the "royal privilege," or monopoly, on calculating machines that Pascal obtained in 1649 suggested that he knew what the Pascaline was up against. The text acknowledges how hard it is to manufacture machines such as the Pascaline and consequently how expensive they must be, and why it is that an inventor (such as, oh, Blaise Pascal, for example) should be given time to refine his device and bring down its cost.[34] In a textbook example of why monopolies are bad for innovation, it appears that Pascal then stopped work on the Pascaline in favor of promoting it to the "influencers" of his day. One of the few known letters Pascal wrote on the subject was sent in 1652 to Queen Christina of Sweden, beseeching her to accept his gift of a clockwork calculating machine.[35]

In the end, Pascal may be forgiven for neglecting his invention. His father died in 1651, leaving only the chronically infirm Blaise and his sister, Jacqueline, in the Parisian home to which the family had returned some years earlier. Within months, Jacqueline had left to become a nun, and, although Pascal continued to correspond with a rarefied group of pen pals (letters he exchanged with Pierre Fermat in the summer of 1654, for example, sowed the seeds of probability theory), he found himself increasingly preoccupied with questions of religion rather than science.[36]

The winter of 1654 was a turning point. In October, Pascal's carriage was left dangling over the Seine when his horses bolted.[37] The following month, still spooked by the experience, Pascal had a religious dream that shook the last remnants of the scientist out of him. He ended his correspondence with Fermat and put aside a work in progress on air pressure.[38] His last major text, published eight years after his death in 1662, was a defense of Christianity in which Pascal laid out his famous wager: if God exists, there is nothing to be lost by believing but everything to be lost by failing to believe.[39] If only Pascal had believed a little more in his own invention.

○——○

THE PASCALINE DID COUNT one notable fan in the form of an Englishman two years Pascal's junior. In 1653, Samuel Morland, a graduate of Magdalene College* in Cambridge, England, was posted to the royal court of Sweden as a diplomat.[40] Morland, having spent the better part of a decade cloistered at Magdalene, was too green to contribute to the negotiations between the English ambassador and Christina, the queen of Sweden. He amused himself during the yearlong mission by writing a textbook on French grammar, debating scholars of Latin, and, it seems, tinkering with a fascinating new adding machine that had been gifted to the queen by a M. Pascal of Paris.[41]

Morland landed a second diplomatic assignment two years later, this time to the Savoy region of France. In neighboring Piedmont, in the spring of 1655, the duke of Savoy had presided over the massacre of a Protestant sect called the Waldensians, and this did not sit well with Britain's newly installed "Lord Protector," Oliver Cromwell, who harbored ambitions of creating a "Protestant League" in Europe, with England at the head of the table.[42] Morland was dispatched to chastise the duke and give English money to the Waldensians and, having done so, he made his way to Geneva to serve under Dr. John Pell, Cromwell's ambassador to Switzerland.[43] (Pell, quite coincidentally, is likely to have created a symbol that adorns pocket calculators the world over. A Swiss student of Pell's, Johann Rahn, wrote an influential math textbook called *Teutsche Algebra* in which he proposed the use of the *obelus*, or "÷", as a symbol for division. Pell is thought to have given him the idea.)[44]

---

* As fans of the British television program *University Challenge* will already know, "Magdalene" is pronounced "Maudlin." The college was named for St. Mary Magdalene but its founder, Lord Audley, insisted on spelling it "Maudleyn," after himself. The pronunciation stuck but the spelling did not.

Upon his return to England, Morland published an anti-Catholic diatribe about his journey, entitled *The History of the Evangelical Churches of the Valleys of Piedmont [. . .] with a most naked and punctual relation of the late bloudy massacre, 1655*. It was partly funded by the deeply Protestant English government and opened with an expansive dedication to England's puritanical, and increasingly dictatorial, Lord Protector.[45] Morland would soon regret it.

Cromwell died suddenly (too suddenly, went the rumors) in 1658, overcome by a combination of malarial symptoms and kidney disease.[46] Within a year, Morland was dismissed from a lucrative position at the Post Office in which he had inspected mail for signs of sedition and made copies of the offending correspondence. By now a seasoned player of diplomatic games, at the same time Morland had been secretly sending intelligence to the king in exile, Charles II. The restored king rewarded Morland with a knighthood in April 1660, but Morland's duplicity raised whispers that he was "a perfidious fellow that betrayed his own masters, and will have spy's reward."[47] Worse, Morland's finances were in disarray: having wheedled a pension from the post office, he promptly sold it to a fellow noble in fear that the institution's debts were about to come due.[48] His resources dwindling, and with Queen Christina's miraculous adding machine at the back of his mind, Morland pivoted to a new enterprise:

> Now finding myself disappointed of all preferment and of any real estate, I betook myself to the mathematics and experiments such as I found pleased the King's fancy.[49]

THE DATES ARE CONFUSED, but it appears that Morland's first attempts to "please the King's fancy" came to fruition between 1662 and 1666. As he described later in a book on the subject, Morland dedicated two devices to the king: a "new and most useful instrument for addition and subtraction of pounds, shillings, pence and farthings; without charging the memory, disturbing the mind or exposing the operator to any uncertainty; which no method hitherto published can justly pretend to"; and a "machine nova cyclologica pro multiplicatione."[50] Or, stripped of Morland's flowery language, an adding machine and a multiplying machine.

Having seen Pascal's adding machine in person all those years ago, Morland would have been keenly aware that there very much *was* a "method hitherto published" for addition and subtraction. He would also have known that his own adding machine paled in comparison with the Pascaline: although it was flat and pocketable, four by three inches in size, its eight dials were independent and had no carry mechanism. Conceptually, Morland's adding device was little better than an abacus or counting board; in practice it was even worse, since its eight dials were labeled and numbered for use with England's archaic coinage, in which a guinea comprised twenty shillings, a shilling twelve pence, and a penny four farthings.[51] This was an accountant's tool, not a mathematician's, and one of scant utility at that. As Robert Hooke, one of the foremost scientist of Morland's day, wrote in his diary:

> Saw Sir S. Morland's Arithmetic engine[.] Very Silly.[52]

Morland's "cyclologica," too, was a letdown. Sized for a desk rather than a pocket, it was essentially a semiautomated version of Napier's bones that required the user to swap numbered disks between axles and add up the resulting partial products, presumably with Morland's own adding machine. Only one example of the former survives, along with four of the latter, suggesting that Morland was no more successful than Pascal in finding a wider audience for his invention.[53]

Despite their obvious shortcomings, Morland's devices found an

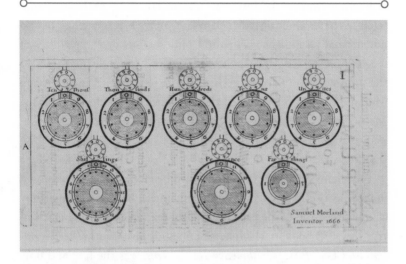

O Illustrations of Sir Samuel Morland's adding machine and "cyclologica,"
or multiplying machine, from the pages of his 1673 book *The Description and
Use of Two Arithmetick Instruments*. Despite their superficial similarity to
Schickard's *Rechenuhr* and Pascal's eponymous Pascaline, Morland's devices
were rudimentary by comparison: his adder could not carry between digits, and
his multiplier did not, in fact, multiply. *Samuel Morland, The Description and Use
of Two Arithmetick Instruments: Together with a Short Treatise, Explaining and
Demonstrating the Ordinary Operations of Arithmetick (London: Moses Pitt, 1673),
figs. 1 and 7. Courtesy of The Linda Hall Library of Science, Engineering & Technology.*

unlikely patron in the form of Cosimo III, duke of Tuscany and scion of the infamous Medici clan. Cosimo was deeply religious and not normally indulgent of natural philosophers or the peddlers of mechanical devices, but having learned of Morland's machines in the summer of 1668, he expressed a desire to acquire copies of them. Morland obliged but declined the duke's repeated offers to join the Medici court. Even in his chronically straitened state, the Englishman was worried that the Catholic Medici would assassinate him should he ever set foot in Florence.[54]

Morland's machines may not have been as groundbreaking or, indeed, as useful as their inventor imagined, but he can rightfully claim authorship of the first meaningful text on computing machinery. The 1673 book in which he describes his inventions, *The Description and Use of Two Arithmetick Instruments*, was the first and only substantial text to be published on the subject before Charles Babbage would revive it in the nineteenth century.[55] For that alone, Morland deserves a place in the annals of computing history.

I T MAY SEEM ODD to dwell on three machines that never fulfilled their promise. Schickard's machines were struck by disaster and lost to obscurity; the Pascaline was confined to the closeted world of the idle rich; and Samuel Morland's devices were overshadowed by his picaresque lifestyle. Indeed, none of these inventors ever seemed to truly believe in their machines. Schickard had his astronomy and mathematics, Morland was fixated on climbing the social ladder, and Pascal had—well, he had anything he turned his mind to. And yet, the act of merely *trying* to build a calculating machine was revolutionary in itself. Without Pascal, Morland, or Schickard, the future of the pocket calculator might have turned out very differently indeed.

5 THE ARITHMOMETER AND THE CURTA

ASCAL'S AND MORLAND'S efforts (Schickard's having been cruelly neglected by history) informed a host of copycat calculating machines. Starting in the eighteenth century, a succession of French, Italian, German, and British inventors created devices of varying efficacy that promised to add, subtract, multiply, or divide as their user required. They presented their machines to royalty to curry favor, to learned societies for their approval, and to the newspapers to gain publicity. None of them managed to make both a successful machine *and* a living out of it, and, as such, the names of Poleni, Lépine, Braun, Stanhope, and others live on as footnotes rather than landmarks in the history of the pocket calculator.

As it turns out, the first truly successful mechanical calculator would also be the progenitor of the last, with a combined lifetime that spanned a century and a half. The machine called the arithmometer, an elegant instrument of brass and wood, was born in Napoleonic France; its descendant, the Curta, resembling nothing so much as the world's most intricate coffee grinder, would be born in a concentration camp. Together, they defined the era of the mechanical calculator and, ultimately, showed the way to the future.

O———O

F OR MUCH OF the eighteenth and nineteenth centuries, the promises made by the inventors of calculating machines and the real-life capabilities of their gadgets were separated by a shimmering gulf of optimism. A host of devices resembling typewriters, cash registers, and clocks in various combinations arrived during that time, yet none of them managed to simultaneously embody practicality, reliability, and affordability. As late as 1872, an American trade publication called *The Manufacturer and Builder* could justifiably lament the state of the art:

> If a reliable calculating machine could be manufactured to retail at a low price, say five dollars, with which addition, subtraction, multiplication, and division could be done, it would no doubt find a ready-sale [. . .] In looking over the models of calculating machines in the Patent Office last month, we saw none that answered the above description. Some of them were so complicated that it would take an engineer to run them, and a watch-maker to keep them in order, while others were evidently designed by men who knew nothing, practically, of the working of machinery [. . .] Finally, every machine was out of order, and gave arithmetical results that would bankrupt the most successful business man in two turns of the handle.[1]

Over in Europe, however, one man was tantalizingly close to making all of this a reality—if at a rather higher price point. His name was Charles Xavier Thomas de Colmar.[2]

Thomas was born in 1785 in Colmar, in the winemaking region of Alsace in eastern France (or in southwestern Germany, depending on the century and/or decade), to a father who practiced medicine and a mother who, somewhat inevitably, has been left out of the history books.[3] Those same history books gloss over Thomas's early life, too, pausing only to mention that he progressed through the ranks of the local administration before joining French troops stationed in the Iberian Peninsula. By 1813, Thomas was responsible for managing provisions for all French forces in Spain, a position that benefited from an accountant's eye and demanded large numbers of calculations. It was here, the legend goes, that he first hatched his idea for the arithmometer.[4]

On returning to civilian life, Thomas founded an insurance company called Le Phénix, or "the Phoenix," which he left after some kind of disagreement. By 1829, he had started another business, Le Soleil, or "The Sun," where things seemed to go rather better. It was at Le Soleil that Thomas pioneered the concept of rolling insurance policies, so that, as a contemporary magazine article explained, *les assurés sont prémunis contre leur propre négligence*—"policyholders are protected against their own negligence."[5] (As a happy coincidence, Le Soleil's bottom line was protected from it too.) In his insurance work, Thomas again found himself in a position where calculation was paramount. He dusted off his idea from Spain and set to work.

At first, Thomas's work proceeded apace. By 1820, he had applied for a patent on a *machine appelée arithmomètre, propre à suppléer à la mémoire et à l'intelligence dans toutes les opérations d'arithmétique*, "a machine called 'arithmometer,' suitable for supplementing memory and intelligence in all arithmetic operations," so that his branding was in order at a very early stage.[6] But to make a machine worthy of the name would take thirty years.

THOMAS'S MACHINE DIFFERED fundamentally from the Pascaline (still, at that time, the most famous almost-success to have gone before) in that it revolved, literally, around a component called a "stepped drum." As shown in the drawing on page 99, this was a cylinder embossed with nine lengthwise splines of increasing length. This cylinder, or drum, meshed with a nine-toothed sprocket that could be slid lengthwise to engage more or fewer of the drum's splines. When the drum revolved around its axis, the sprocket would rotate through a smaller or larger arc, depending on the number of engaged teeth. For instance, if the sprocket was slid right down to one end of the drum, where only one spline remained, the sprocket would rotate by one tooth only for each revolution of the drum. If the sprocket was slid upwards a little, so that two of the drum's splines engaged it, a single turn of the drum would rotate the sprocket by two teeth. And so on, all the way up to the top of the drum, where all nine splines would drive the sprocket round a full revolution for each turn of the drum.

Thomas's arithmometer used one stepped drum for each input digit, and each drum's sprocket was connected in turn to a simple counting mechanism. Sprockets were positioned by means of sliders labeled from 0 to 9, so that the user could see precisely how many teeth would be engaged by each turn of the drum. Set a slider to 2, for instance, and each turn of the drum would add 2 to the corresponding counter. Turned once, the counter went from 0 to 2; turned again, from 2 to 4; a third time, from 4 to 6; and so on. Cleverly, the stepped drum worked just as well in reverse: by turning a drum in the opposite direction, the currently set number was subtracted from the counter, not added.[7]

The stepped drum was not a new invention. It was the brainchild of a mathematician and philosopher named Gottfried Wilhelm Leibniz, a man so argumentative that only his secretary attended his funeral, and whose philosophical assertion that we must necessarily live in "the best of all possible worlds" was mercilessly satirized by, among others, the celebrated French writer Voltaire.[8] For all his faults, Leibniz was an indefatigable scholar, vaulting between disciplines as he pleased. He created binary arithmetic, inspired by an ancient Chinese book called the *Yì jīng* (often spelled "I Ching").[9]

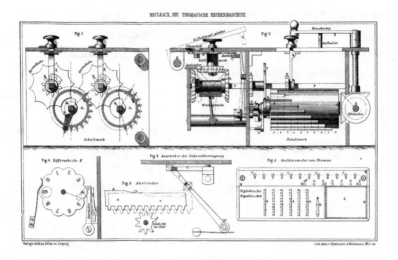

REULEAUX, DIE THOMAS'SCHE RECHENMASCHINE

O An 1862 drawing of the internals of Charles Xavier Thomas's arithmometer.
Fig. 2 shows a section through one of the device's counters. Visible are the
stepped drum that drives the counting mechanism, the slider that selects
how many of the drum's splines will engage the counting mechanism during a
rotation, and the counting mechanism itself. *Reuleaux, Franz. Die Sogenannte
Thoma'sche Rechenmaschine (Leipzig: Felix, 1892), 61, http://resolver.sub.
uni-goettingen.de/purl?PPN599461640. Public domain image courtesy of
Niedersächsische Staats- und Universitätsbibliothek Göttingen.*

And he would later be given dual credit, with Isaac Newton, for having invented "calculus," the mathematics of gradients, change, integration, and differentiation, whose name descends from the Roman
counting tokens called *calculi*.[10]

In 1672, Leibniz got the opportunity of a lifetime. After years of
trying to break into Europe's intellectual circles, he was sent to Paris
on a diplomatic mission for the Elector of Mainz, one of seven dig-
nitaries who elected the Holy Roman Emperor.[11] During four years
spent in France, Leibniz met some of the continent's most famous
thinkers, and, moreover, gained access to some of Blaise Pascal's
unpublished writings, which may have introduced him to the idea

of a calculating machine.[12] His ambition was further kindled when he saw a Pascaline in person in 1672, and then, a year later, when he was shown Samuel Morland's machines during a trip to London.[13] If a human could imbue a machine with the rules of mathematics, the pious Leibniz mused, could God not have imbued humanity with life itself?[14]

Despite his best efforts, Leibniz's "stepped reckoner" device was never quite good enough. He demonstrated a wooden version at the Royal Society in London and was invited to return when it actually worked. Back in Paris, he explained the problems with the machine to a clockmaker named M. Olivier and left Olivier to fix them. The finished machine, rediscovered in an attic at Göttingen University in Germany at the tail end of the nineteenth century, still did not work.[15] Nevertheless, the stepped drum at the heart of Leibniz's calculator was an inspired solution to at least some of the problems that Schickard, Pascal, and Morland had faced. All it needed was the right person to put it into action.

Charles Xavier Thomas was that person. It is possible that he arrived at the concept of the stepped drum with no outside help, but scholars have noted that descriptions of Leibniz's stepped drum had been publicized as early as 1744, so that, had Thomas gone looking for inspiration, he may well have found it.[16] Regardless, Thomas put the stepped drum to better use than anyone before him.

THOMAS ITERATED THROUGH so many arithmometers that they are better approached as a general class of machines rather than individual models. Most are similar in appearance and operation, sized to fit comfortably on an office desk and made up of two distinct parts. The lower, fixed part of the arithmometer holds the machine's "inputs": a set of vertical sliders, each labeled from 0 to 9, that together represent a number to be added to or subtracted from the running total. Behind each slider is a stepped drum and sprocket; a counting mechanism; and a carry mechanism that carefully distributes carries one at a time, thereby avoiding the problems had plagued Schickard's and Pascal's machines. All of the drums in the lower half of the machine are driven by a single common crankshaft and handle: turn it once to add the inputs to the running total; turn it again to re-add them, and so on.

The movable upper part of Thomas's device does nothing but record the running total. It holds a series of numbered dials driven by the mechanisms in the lower part of the machine. Like a typewriter

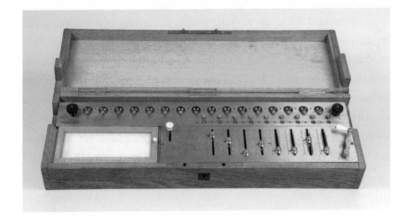

O An arithmometer made around 1860, some forty years after Thomas first patented his device. This model was used at the Blue Hill Meteorological Observatory in Milton, Massachusetts. Input numbers, set by the sliders at bottom, can be up to eight digits long, and the sixteen counters on the movable carriage at top supported running totals of up to twice that length. Each turn of the crank handle to the right added or subtracted the current inputs to the values in the carriage. *Thomas de Colmar, Charles Xavier. "Arithmometer (1990-5-0026)." Collection of Historical Scientific Instruments. Cambridge, MA: Harvard University, 1860. http://waywiser.fas.harvard.edu/objects/3894/arithmometer. Image courtesy of Collection of Historical Scientific Instruments, Harvard University.*

carriage, the upper part can be moved left or right with respect to the lower part, a feature that allows savvy users to carry out large multiplications with ease. Imagine, for example, we need to multiply 234 by 10. We *could* set 234 as an input number and then turn the crank ten times, but that is a lot of cranking. The smart thing to do is shift the carriage right by one position, so that the units slider (set to 4, in this case) now goes into the tens dial, the tens slider (set to 3) goes into the hundreds dial, and the hundreds slider (set to 2) goes into the thousands dial. Now, a single turn of the handle adds not 234 to the output dials but rather 2,340. (More complex multiplications are possible too.) It is a neat optimization for the problem of large multiplications, and even today some microprocessors still multiply using similar shift-and-add methods.[17]

There were other clever features too. A counter beneath each dial in the carriage showed how many times the crank had been turned, so that the user could track their progress through complex shift-and-add multiplications such as the one described above. Subtraction and division were possible, too, simply by moving a switch that reversed the direction in which the crank turned the sprockets. Accidentally turned the crank once too often during a multiplication? Switch to subtraction mode and crank it one more time to remove the excess.[18]

HAVING FILED his first patent in 1820, Thomas submitted an improved machine a year later to the Société d'encouragement pour l'industrie nationale, a body set up to promote French industry, which issued glowing reports on the arithmometer's design and function.[19] That same year, Thomas was knighted for his work on the machine—and then did little or nothing with it for the next twenty years.[20] The interregnum may have been a product of his demanding insurance work; or, as Stephen Johnston, writer of a clear-eyed history of Thomas's machine, has it, the arithmometer may not have been the ready-made miracle that the Société painted it to be.[21]

Whatever the reason, two and a half decades after starting work on the arithmometer, Thomas was finally ready to declare it finished. In the wake of the French Revolution at the end of the eighteenth century, republican leaders had staged a series of national *fêtes*, or festivals, which, by the time Thomas found himself searching for a platform to launch his finished invention, had evolved into showcases for French industry and culture. (The staging of London's Great Exhibition of 1851 was the sincerest form of flattery, and the concept of the national exhibition, or expo, is still alive today.)[22] Thus it was that the arithmometer debuted in Paris in 1844, one of a host of "diverse measures, counters and calculating machines," only to be overlooked for a medal in favor of a competing machine judged to be the more capable. Thomas lost out again in 1849, this time to a rival called the *arithmaurel*.[23] Two years later, at the Great Exhibition in London, Thomas's machine was honored by inclusion in the event's catalog—but to the chagrin of its inventor, it was another entrant, a machine made by a Polish watchmaker, that caught the eye of the press.[24]

In Paris once again for the Universal Exposition of 1855, Thomas swore he would not be upstaged. Judging the arithmaurel to be his main competitor, he made a giant arithmometer six feet long, with fifteen input sliders and thirty result dials allowing results as large as 999,999,999,999,999,999,999,999,999,999, and installed it in a grand wooden cabinet. To make his case in advance of the event, he contrived to have published a hundred-page celebration of the arithmometer that masqueraded under the innocuous title of *Histoire des nombres et de la numération mécanique* (History of numbers and mechanical numeration). Ever the bridesmaid, the arithmometer was overshadowed yet again, but not by the arithmaurel. Rather, it was a late entry to the exhibition, a so-called difference engine designed in Sweden, which could calculate and print tables of logarithms to five decimal places, that took the gold medal.[25] The arithmometer, with its addition, subtraction, multiplication, and division, seemed rudimentary by comparison.

Thomas's disappointment at yet another snub would surely have been blunted by the five additional knighthoods he had been awarded in 1852, 1853, and 1854; by the many gifts and testimonials bestowed on him by the great and the good of France; and by the growing fortune amassed from his insurance company, Le Soleil. When he died in 1870 at one of his many Parisian properties (having added yet another knighthood to his name in the interim), Thomas left behind an estate worth 24 million francs.[26]

The arithmometer, too, bounced back. Between 1851, when production began in earnest, and Thomas's death two decades later, the workshop he had established in Paris manufactured some eight hundred examples. Production carried on under his successors at Le Soleil, and, for a few rosy, uncontested years, the arithmometer was the undisputed champion among calculating machines.[27] A number of influential customers recognized it as such: Henry Brunel, of the English engineering dynasty, bought an arithmometer in 1872 for £12 and was much pleased with it:

> I have just got what my mother irreverently calls "a new toy"—to wit a calculating machine price £12 which does all the common operations of arithmetic viz addition, multiplication, subtraction & division in the twinkling of an eye. It is really a very useful article worth its weight in brass.*

Brunel would go on to become a formidable booster for the arithmometer in England.[28] Also in England, the Prudential insurance company was a prominent early adopter. When conducting a five-yearly audit in

---

* At the time of writing, the £12 that Henry Brunel spent on his arithmometer equates to around £1,388, or $1,958, in today's money. Not cheap, but broadly in line with the personal computers that populate offices and homes the world over.

1877, the company used twenty-four arithmometers to perform more than 4.6 million calculations over a period of six months. Without them, the company's director said, the audit would have been impossible.[29]

Thomas did not live to see his invention realize its full potential, but even during his lifetime the arithmometer had become the Western world's best known and, better still, most trusted mechanical calculator. It would not fit in your pocket—not yet, anyway—but the arithmometer had proved that a machine could lighten the mathematical burden levied by an increasingly industrialized and connected world.

COURTESY OF an extraordinary interview conducted a few years before his death, we have an unparalleled insight into the life and work of one Curt Herzstark, born in Vienna, Austria, in 1902 to a Jewish father and a Lutheran mother. Herzstark grew up in the world that the arithmometer had built—his father, Samuel, founded a business in 1905 to build electrically powered arithmometers—and Herzstark would, in time, become as pivotal to the arithmometer's history as Thomas himself.[30]

Herzstark's initiation into the mechanical arts started early. Having completed his studies at a specialized engineering high school, he apprenticed for a year in the family factory before traveling to Germany to intern at one of the company's suppliers. There, Herzstark learned how to fix "ten-key adders," simple, key-driven adding machines based on an American design from the turn of the century.[31] Soon enough, he was on the road as a salesman, behind the wheel of an imported Chrysler with an adding machine installed in the backseat for demonstration to potential customers. And as he talked to those customers, he started to understand something about the calculator:

Something was missing from the world market. "I would like to have a machine that fits into my pocket and can calculate. I am a building foreman. I am an architect. I am a customs officer. I have to be able to pick something up. I cannot go ten kilometers to use a calculator in the office. The slide rules are not useful for my purpose. Slide rules cannot add or subtract. And aside from that you can only read three values from the markings on them, not more. For an invoice I have to know exactly."[32]

Not for nothing was the average building foreman or traveling accountant pining for a more portable calculator. A representative "lightweight" model, the Marchant calculating machine of 1935, weighed the same as a four-year-old child and had to be lugged around in a suitcase.[33]

By 1937, Herzstark had devised a plan to exploit this underserved niche. The Herzstark factory would build, he proposed, a miniaturized mechanical calculator—an arithmometer small enough to fit in the palm of the hand but capable enough to replace a desktop machine. New aluminum and magnesium alloys, more commonly used to build airplanes, would make it light and strong; for now, though, to prove that it could work, Herzstark made a handful of Bakelite prototypes.[34]

Reality intruded. Herzstark's father had been easing himself out of the family business for a number of years,* and it was understood that Curt would be named as the factory's owner when the time came. But when Samuel died in 1937 and the legal arrangements dragged on into 1938, Europe's febrile politics upset the family's plans. In March 1938, Austria's fascist-but-not-fascist-enough-for-the-Nazis

---

* In 1930, tired of overseeing the factory, Samuel Herzstark bought a movie theater in Vienna's Prater amusement park. The park is also home to the Ferris wheel on which Orson Welles delivers his infamous "cuckoo clock" speech as part of Carol Reed's film *The Third Man*.

chancellor, Kurt von Schuschnigg, was forced to resign and the country was annexed by German troops.[35] Curt, the son of a Jewish father, would have to quietly run things as an employee of the Herzstark family firm while his Christian mother served as its legal owner.[36]

Gradually, the screws tightened. The occupiers barred the factory from making calculators, to avoid competing with Germany's own manufacturers, then extended an offer the Herzstarks could not refuse: a contract to make measuring gauges for building German tanks.[37] Then, in 1943, a pair of Herzstark employees were caught transcribing English-language radio broadcasts. One was imprisoned, the other executed, and, when Herzstark protested to the Gestapo, he was arrested for his trouble. In short order, he was sent to prison and then on to Buchenwald, the first and largest of Nazi Germany's network of concentration camps.[38]

As a skilled engineer, Herzstark was offered a second Faustian bargain. In exchange for supervising a cadre of inmates who fabricated components for V2 rockets and repaired looted calculating machines, he would be allowed to live. He used his position to mastermind a number of small rebellions, requesting the transfer of vulnerable prisoners into the camp's machine shop and teaching them how to look useful should an SS guard pass by.[39]

Even in such precarious circumstances, with bombing raids and executions mere facts of life, Herzstark could not contain his enthusiasm for his miniature arithmometer. Having talked to others in the camp's machine shop about his idea, the factory's manager, a German engineer who knew Herzstark from his old life, told Curt:

> I understand you've been working on a new thing, a small calculating machine. We will allow you to make and draw everything. If it is really worth something, then we will give it to the Fuhrer as a present after we win the war. Then, surely, you will be made an Aryan.

Herzstark realized that his invention could be his salvation. "I thought to myself, my God, if you do this, you can extend your life. And then and there I started to draw the Curta, the way I had imagined it." The blueprints of Herzstark's miniaturized arithmometer, drawn in the evenings and on Sundays, were almost complete when, in April 1945, Buchenwald was liberated. Herzstark was free, finally, to see his idea to fruition.[40]

MOST OBSERVERS peg Herzstark's magnum opus as resembling either a pepper mill or a coffee grinder. William Gibson, in his 2003 novel *Pattern Recognition*, has a more apt description: he calls it a "math grenade." After all, the Curta, which serves admirably as a plot device in that book, was nothing if not a shock to the established order of clunky desktop calculators.[41]

Named to suggest that the device was "Curt's daughter," the Curta was an aluminum cylinder around two and a half inches in diameter and twice as tall, with numbered sliders arrayed around its circumference and a crank handle on top.[42] Herzstark's drawings showed that it worked in the same way as a conventional arithmometer: use the sliders to set an input value, then turn the handle once to add that value to a set of accumulator dials visible on top of the machine. As with Thomas's version, those accumulators could be realigned to multiply the input value by a factor of 10, 100, 1,000, and so on. In lieu of a switch to toggle between addition and subtraction modes, the Curta's user merely had to pull out the handle by a hair and then crank it as normal.[43]

Herzstark claimed to have devised the Curta's appearance and controls—its "user interface," so to speak—before he ever considered its internal mechanisms.[44] Being cylindrical, it fitted neatly into one hand, which could be used to rotate the carriage with the thumb and

O The Curta Type I: the first and last practical mechanical pocket calculator. *CC BY-SA 2.0 image courtesy of Magnus Hagdorn.*

forefinger while the other hand was free to adjust the input sliders and turn the crank. The accumulators and crank rotation counter were located on the top of the cylinder and could be read as the user turned the handle.[45]

Yet if Herzstark had designed the outside of the Curta before its insides, he could hardly have constructed a better shell in which to wrap them. Like Thomas's arithmometer, the Curta relied on the concept of a stepped drum driving a sprocket for each input slider. But rather than the multitude of stepped drums in the original, one for each input slider, the Curta had only one large drum, right in the center of the machine. As the user turned the crank, the splines of the drum engaged with each input sprocket in turn. Imagine, for example, that we have a Curta, freshly zeroed, and awaiting our inputs. We set the first three sliders to 1, 2, and 3 respectively, and slowly turn the crank. A "1" pops into the first dial on the top of the Curta, then, a split second later, a "2" follows in the second dial, and finally a "3" appears in the third, as the splines on the drum sweep past each input sprocket. Our addition is complete. This ingenious packaging allowed the Curta to shrink substantially when compared to its nineteenth-century predecessor yet still support the same arithmetical operations.[46] It was the embodiment of the idea that perfection is not when there is nothing left to add but when there is nothing left to take away.

EAGER TO PUT his plans into action, Herzstark first went to the city of Weimar, less than a day from the Buchenwald camp, where he knew of a Herzstark sales agent from before the war. He was greeted by the city's German inhabitants with sheepish deference: "Excuse us, we haven't known anything for years," he was told, as people on the street moved meekly from his path. He replied, with quite some equanimity, "Thank god, I am alive, you are alive. I am not a judge. Others should take care of that."[47]

Herr Müller, the salesman, was similarly contrite, but Herzstark brushed off his apologies. Why fret about the past, horrific as it was, when there was business to be done? Müller duly introduced Herzstark to the representatives of an engineering firm in nearby Sommertal who, in turn, helped Herzstark manufacture three new prototypes.[48] Then, with rumors circulating that Russia, in whose sphere of influence Weimar now lay, was press-ganging engineers into careers in more easterly climes, Herzstark put the prototypes in his bag and left for Vienna with alacrity.[49]

After a joyful reunion with his family in Vienna, Herzstark again pushed on with the business of making the Curta. But there were problems. His brother, Ernst, who had taken over the factory after Curt's arrest, had no plans to give it up. Marie, his mother, was adamant that the brothers must share any profits that might arise from manufacturing the Curta.[50] Nor was there a surfeit of interest from other quarters: Europe's governments, exhausted by war, did not care greatly to get involved.[51] Except one, that is.

Consequently, in 1946, Herzstark found himself in the Principality of Liechtenstein, an impossibly tiny alpine nation wedged between Switzerland and Austria, overseeing the conversion of a hotel ballroom into a makeshift factory for the Curta Type 1 calculator. Prince Francis Joseph II had offered Herzstark one-third of the stock of Contina AG, a company created solely to manufacture the Curta, in exchange for the chance to drag his rural fiefdom into the twentieth century. Herzstark jumped at the opportunity to become, as he put it, the "Messiah from Liechtenstein."[52]

But it did not go well. Herzstark fought constantly with Contina management, who held sway over all recruitment and expenses.[53] An order for ten thousand Curtas from an American department store was declined because executives felt the company could not produce enough machines.[54] (They may have had a point: in 1949, production was only around three hundred units per month.)[55] Worse yet, in 1950 it emerged that Contina AG was drowning in debt. The old business was dissolved and a new one constituted in its stead so that Herzstark's shares evaporated overnight.[56] A disillusioned Herzstark left the company in 1952, pausing only to wrestle a license fee out of Contina for the patents still held in his name.[57]

The Curta would never sell in bulk. Even after 1952, by which time Contina was making a thousand Curtas a month, its calculators were available only via mail order and in a handful of specialty stores.[58] Even with wider distribution, the Curta's price tag might have put off many potential buyers: an advert in *Popular Mechanics* offered a Curta for $134.70, or almost $1,400 in today's money— which, coincidentally enough, is more or less what a vintage Curta fetches on today's robust secondhand market.[59]

Production came to an end in 1972 after around 150,000 Curtas had been made.[60] A creditable number, especially when Charles

Xavier Thomas's company managed only five thousand arithmometers in total, even counting those made after his death, but Herzstark's intricate, ingenious device never scaled the heights that its inventor had imagined for it.[61] The Curta had dedicated cohorts of followers, certainly: some, such as accountants, were to be expected; others, such as rally drivers, who used Curtas to compute times and distances on the road, less so.[62]

Make no mistake, the Curta was the first calculator that managed to be both portable and practical at the same time—a so-called four-function machine capable of addition, subtraction, multiplication, and division that also happened to fit in one's pocket. But it lay at the end of an evolutionary road, not at the start, and even its inventor seemed to be at a loss as to where to go next. Blueprints and patent drawings discovered in 2017 showed that Herzstark, in his wilderness years after leaving Contina, had considered combining multiple Curtas to carry out several operations at once. His diagrams show a collection of unwieldy, inelegant devices that never came to pass.[63]

Wars like the one that Curt Herzstark suffered through have a way of focusing the collective mind. The development of the Curta, for instance, was driven by Herzstark's desperate efforts to survive the Second World War even as millions of others perished. The same conflict would ultimately give rise to a new type of pocket calculator, although it would take decades for the lessons learned at the Manhattan Project in the United States, at Bletchley Park in England, and at other secretive projects around the world to come to fruition. In the meantime, the mechanical calculator was going nowhere.

6 THE FRIDEN STW-10

OR THE LUCKY FEW who owned one, the Curta was a marvel both at rest and in use—an ingenious device, brilliantly executed, that liberated its user from the office desk and the typewriter-sized calculator that dwelt upon it. But the jewel-like Curta was a rarity, representative of calculators only in the same way that a Ferrari is representative of cars. Far more common were the utilitarian, battleship gray calculators that graced the desks of countless office workers. Many were made by stolid, familiar corporations that had become part of the business landscape: Smith Corona, best known for its typewriters; Burroughs, an adding machine pioneer founded in 1886; and Monroe, a maker of calculators based on an expired nineteenth-century patent.[1] These were unexceptional machines for the most part, little advanced from the electrically driven arithmometers first built at the turn of the century, and the vast majority were capable only of the four fundamental arithmetical operations.[2] And yet, by the mid-twentieth century, when the Curta was garnering attention but not sales, those same simple calculators had become the hardware on which ran great enterprises of commerce and science. They were the hardware and we, their human wranglers, were the software.

O———O

ALAN TURING, the twentieth-century British mathematician, is known for many things. At Bletchley Park, a stately home that housed Britain's government codebreakers, he masterminded the cracking of the German "Enigma" encryption scheme, a feat that may have shortened the Second World War by up to two years.[3] Before the war, Turing had published a conceptual blueprint for all programmable electronic computers, later to be dubbed the "universal Turing machine." After the war, he devised the "imitation game," or Turing test, that anticipated the arrival of artificial intelligence.[4] Finally, Turing is remembered for the manner of his death. He was convicted of gross indecency in 1952 for a relationship with another man and was offered estrogen injections, a form of "chemical castration," to avoid prison. He endured the mood swings and depression caused by the drug only to die two years later of cyanide poisoning. Whether he took his own life remains an open question; either way, a posthumous pardon granted in 2013 seems scant compensation for a grossly indecent punishment.[5]

For much of his lifetime, Turing would have had a quite different understanding of the word "computer" as compared to its current meaning. In the 1950 paper in which Turing explains the imitation game, he describes the job of what he calls a "human computer":

> The human computer is supposed to be following fixed rules; he has no authority to deviate from them in any detail. We may suppose that these rules are supplied in a book, which is altered whenever he is put on to a new job. He has also an unlimited supply of paper on which he does his calculations. He may also do his multiplications and additions on a "desk machine," but this is not important.[6]

Turing used the term only because he was writing specifically about whether an *electronic* computer could be considered to be a thinking machine. For Turing, along with most other mathematicians, scientists, and economists of his day, the word "computer" did not ordinarily require a qualifier: for centuries, computers *were* humans, workers on a mathematical assembly line whose job it was to carry out meticulous, repetitive calculations in the service of some greater goal. There is an argument, if a tenuous one, that computing has existed for millennia: at least one Roman gravestone depicts a scene in which the deceased calls out numerical amounts and a servant records them with his counting tokens.[7] A similar activity is shown in a medieval book on arithmetic. Here, three seated abacists manipulate numbers dictated by two others.[8] As with many significant events in history, however, the advent of the modern practice of human computation is tied to a portent from the heavens: the return, in 1758, of Halley's Comet.

N THE SEVENTEENTH CENTURY, England's Astronomer Royal, Edmond Halley, had tried and failed to factor in the gravitational forces of Jupiter and Saturn as he worked to predict when his eponymous comet would return.[9] He figured that the comet must have an orbital period of between seventy-five and seventy-six years, but to be more precise than that was beyond him, thanks to a conundrum called the three-body problem. The problem is this: it is almost impossible to write down, using conventional mathematics, the behavior of any real-life collection of three or more bodies that are gravitationally attracted to one another—say, the sun, Jupiter, Saturn, and Halley's Comet. The motion of a two-body system is predictable, but add a third, and all bets are off.[10]

By the middle of the following century, however, a few years shy of the comet's predicted return, a French astronomer named Alexis Clairaut had come up with a novel way to solve the three-body problem. Rather than try to collapse all of the factors governing the comet's motion into a single equation, as Halley had done, Clairaut proposed instead to follow the comet around the sun in incremental steps. Starting with a recent observation of the comet, he would calculate the gravitational pulls of both Saturn and Jupiter and figure out their effects on the comet at that precise moment. Having advanced the comet along its elliptical path for a degree or two, he would recalculate those same gravitational pulls at its new location and repeat the process. And so on, and so on, until the comet had reached its perihelion, or closest approach to the sun.[11]

Clairaut's method was easy to grasp, but as we might say today, it was computationally expensive. He recruited another astronomer, Joseph Jérôme Lalande, and one Nicole-Reine Lepaute, the wife of the French royal clockmaker, to help him work toward his goal. Lalande and Lepaute computed planetary positions and gravitational forces, which Clairaut then used to advance the comet along its simulated path. Both their divide-and-conquer strategy and Clairaut's use of numbers rather than algebra were uncannily predictive of how electronic calculators and computers would eventually solve similar problems.

It took the trio six grueling months to arrive at a tentative conclusion, and late in 1758, with the comet's return imminent, Clairaut announced their results.[12] In November that year, he told the Paris Academy of Sciences that Halley's Comet would reach its perihelion on April 15, 1759. The observed date was March 13. His estimate was more accurate than Halley's by an order of magnitude, making Clairaut a celebrity and prompting suggestions that the comet should be renamed after Clairaut, its oracle, rather than Halley, its discoverer.[13] That did not happen, of course, and Clairaut preferred to dine out on his newfound fame than to pursue his scientific career, but his feat is remembered as the first formal experiment in human computation.[14]

○———○

SCIENTISTS ACROSS EUROPE rushed to apply Clairaut's methods to their work. The USA, by contrast, was less enthused. American astronomers and mathematicians mostly worked alone, a problem compounded by a general distrust of government that, in turn, made impossible the kind of state sponsorship that had underwritten many of Europe's great scientific discoveries. But all this would change. The USA was growing, and modernizing, very quickly indeed.[15]

In 1849, the U.S. Navy established the American Nautical Almanac Office, an almanac being a book of tables and other observations that can be used to navigate by the sun, moon, stars, and planets. Charles Henry Davis, the captain in charge of the office, pushed for it to be sited in Cambridge, Massachusetts: for one, there was a ready supply of scholars from nearby Harvard College; for another, he already lived in that same city.[16]

Davis's first order of business was to decide where to place the almanac's prime meridian. All locations on Earth, or in the sky as observed from the Earth, can be pinpointed using a latitude and a longitude. Both are angular measurements taken relative to a fixed line: for latitude, that line is the equator; for longitude, it is a "prime meridian" that runs from pole to pole through some fixed geographical point. British almanacs used a prime meridian running through the Royal Observatory in Greenwich, a London suburb that overlooked the River Thames, but Davis wanted a meridian on U.S. soil. The problem was that American sailors had no incentive to abandon the Greenwich meridian, which was familiar to them from their existing almanacs, and so Davis proposed a compromise. The U.S. almanac would come in two volumes: one, with measurements made relative to Greenwich, would be aimed at navigators at sea; the other, for astronomers and surveyors, would use an American prime meridian running through the city of New Orleans.[17] It was an exceptionally neat arrangement. New Orleans was exactly 90 degrees from Greenwich, or one-quarter of the way around the globe, which made it easy to transpose longitudes from one meridian to the other.[18]

The powers that be in Washington and New York, many thousands of miles away from New Orleans, were not persuaded. Davis was instructed by Congress to use a meridian fixed at the U.S. Naval Observatory itself, sited on the banks of the Potomac River in emulation of its counterpart in Greenwich.[19] With his dual meridians now decided and many thousands of calculations yet to be made, Davis chose to further imitate the Greenwich observatory. Since 1835, George Airy, one of Edmond Halley's successors as Astronomer Royal, had been crunching astronomical data with the aid of a battalion of human computers. Airy's computers, all of them male and some as young as fifteen, routinely worked twelve-hour days in the observatory's octagonal office. A rule stated that any computer who reached the age of twenty-three would be summarily dismissed.[20]

Davis organized his computers with similar discipline but was less dogmatic in recruiting and retaining them. Foremost among his computers was Maria Mitchell, a skilled astronomer and mathematician born in 1818.[21] Mitchell, the only woman on the team, had already received a medal from the king of Denmark for identifying a new comet (one of the USA's first major scientific discoveries) and would later be appointed the first professor of astronomy at Vassar College.[22] Mitchell, along with half of her co-computers at the almanac office, corresponded with Davis via mail; computing was still not an established profession, and permanent computing offices were scarce.[23]

Somewhat predictably, Davis assigned Mitchell to process observations of Venus: "As it is 'Venus who brings everything that's fair,'" he wrote, "I therefore assign you the ephemeris [trajectory] of Venus, you being my only fair assistant."[24] One can almost imagine Mitchell's eyes rolling. As she wrote later, "The eye that directs a needle in the delicate meshes of embroidery will equally well bisect a star with the spider web of the micrometer."[25]

Maria Mitchell blazed a trail for other human computers. Harvard College Observatory, also in Cambridge, hired its first female computer in 1875, and by 1880, the entire computing staff was made up of women. But this was not quite the victory for equality that it appeared to be. Harvard's female computers were paid only half the wages of their male predecessors.[26]

WORLD WAR I brought the need for computing into sharp focus. U.S. Army computers, many of them college-educated women, produced map grids, surveying aids, and gunnery tables.[27] (One of them, Elizabeth Webb Wilson, later recalled her discovery that "the Germans had the advantage because the earth turned towards the east, therefore as they were shooting towards the west, their bullets travelled further into the allies' lines.")[28] Similarly, Herbert Hoover's Food Administration used computers to predict America's wartime agricultural needs.[29]

Between the wars, yet more institutions edged gingerly into the waters. The U.S. Department of Agriculture took on computers to process farming data. AT&T's staff of eleven computers tackled scientific work. Colleges and universities used computation to solve thorny numerical problems and produce mathematical tables.[30] And in 1938, in New York City, a project began that would professionalize the industry and, at the same time, suffer acutely from the frailty of its human components. Led by Gertrude Blanch, a Polish immigrant to the USA, the Mathematical Tables Project would be one of the most enduring and successful efforts to mold hesitant human brains into a numerical production line.

The MTP was part of a lingering countrywide hangover from the Great Depression. Established by the Works Progress Administration of the federal government, the MTP's mandate was to compute and publish mathematical tables while, at the same time, employing as many workers as possible.[31] Gertrude Blanch, who funded her first degree in mathematics and physics by working for a milliner and who had been pursued by Cornell University for her PhD, was brought in to oversee the work.[32]

By the spring of 1938, the MTP had recruited 125 computers, making it one of the largest computing groups in the world, although many of its reluctant employees languished in states of physical distress or mental ill health.[33] Some did not understand arithmetic. Others pushed for unionization, and yet others deliberately sabotaged their work. It was not, on the whole, a happy place. And yet, Blanch persisted. By 1941, the project had taken on a staggering 450 computers and had published twenty-eight volumes of tables: page after page of sines, cosines, logarithms, exponentials, probabilities, and many other exotic mathematical functions.[34]

Blanch took Alexis Clairaut's divide-and-conquer technique to new levels. She organized her pen-and-paper computers into different groups for different arithmetical operations: one group carried out only additions, another subtractions; a third performed multiplications by a single digit, and a fourth, comprising the most able workers, handled long divisions.[35]

With echoes of other WPA projects, which favored shovels over power tools and wheelbarrows over trucks, the MTP was required to use labor-intensive methods wherever practical.[36] The few computers lucky enough to have a mechanical calculator on their desk would start their day by punching in a few random numbers and cranking the handle to find out if their machine had any maladjusted levers or sticking gears. By the end of the day, their arms would ache from having pulled that same lever over and over again.[37]

○———○

SCALE ASIDE, the Mathematical Tables Project was an out-
lier in another way. Under WPA rules, no more than 1 in 5
of its employees could be women.[38] Over at NACA, however, the
National Advisory Committee for Aeronautics, there were no such
restrictions. NACA was free to populate its own computer rooms
with women—and to discriminate against them as it saw fit.[39] The
organization's mostly male, college-educated engineers knew that
computation was essential to their jobs, and they knew also that
NACA's mostly female, college-educated computers were simply better at it than they were, but the men appealed to their bosses' sense
of manly pride to excuse themselves from what they saw as a menial
task.[40] From a memo of 1942:

> The engineers admit themselves that the girl computers do the
> work more rapidly and accurately than they would. This is due
> in large measure to the feeling among the engineers that their
> college and industrial experience is being wasted and thwarted
> by mere repetitive calculations.[41]

Attitudes like this were common. On the other side of the Atlantic, a British physicist and eventual Nobel laureate named Cecil Powell asked a colleague for "three more microscopes and three girls" to
help analyze photographs of atomic processes.[42] And it was without
irony that a member of the U.S. National Defense Research Committee could, in 1944, refer to a unit of computing power as a "kilogirl."[43] Just as today the word "computer" puts us in mind of a faceless
electronic device, for the male scientists of the era, a computer was an
organic, female version of the same thing.

At least NACA's computers were better equipped than their
MTP counterparts. A typical job, in those early days, was to take

wind-speed measurements from anemometers inside wind tunnels and then average out and plot the results for use by NACA's engineers. One-thousandth of a kilogirl might, at her desk, use a push-button adding machine called a comptometer, an electrically driven mechanical calculator, or, in a pinch, a large, double-sided slide rule.[44] The same basic equipment was still in use in 1962, as the rechristened National Aeronautical and Space Administration prepared to launch John Glenn on America's first crewed orbital spaceflight.

Codenamed Friendship 7, the mission was already two years late. Moreover, Glenn was skeptical of the IBM 7090 mainframe that had been used to calculate his rocket's trajectory. A test pilot by trade, Glenn was more trusting of people and processes, which he could understand, than he was of electronics, which he did not. "Get the girl to check the numbers," he pleaded after a final ride in the capsule's ground-based simulator.

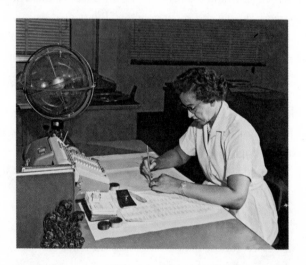

O   Katherine Johnson at work in 1962. On her desk is a Monroe mechanical calculator, rather than the Friden she would use to help John Glenn orbit the Earth and return safely. Nye, Bob. "Katherine G. Johnson at Work." NASA Images, 1962. https://images.nasa.gov/details-LRC-1962-B701_P-09381.

The "girl" was one Katherine Johnson, a forty-three-year-old Black woman, and she was ideally placed to give Glenn the reassurance he needed. In her nine years at NASA, Johnson had coauthored a paper on orbital mechanics that informed, in part, the computer programs that Glenn did not trust.[45] With three days to spare before the launch and an astronaut needing reassurance, Johnson sat down at her electromechanical Friden STW-10 calculator and got to work.[46]

IN ALL WAYS, the machine that Katherine Johnson would rely upon to put a man in orbit was robustly unexceptional. The Friden model STW, unveiled in the fall of 1949 at the New York Business Show, was a gray wedge slightly more than a foot wide with an abundant keyboard and a "carriage" that jutted out by six inches.[47] It weighed about forty pounds, sported buttons for addition, subtraction, multiplication, and division, and needed an electrical socket for power. Its debut, most likely, would have raised few eyebrows. After all, barring a few changes to its casing and color scheme, the STW was almost identical to its predecessor, Friden's venerable ST, which had been in production since 1936.[48]

Internally, both the ST and the STW used the arithmometer's stepped drums and counting mechanisms; externally, they shared its movable carriage. That carriage, however, could move itself automatically from digit to digit during large multiplications and divisions, saving the user from having to do so themselves. This was not, it should be said, a cutting-edge feature: the first "automatic" arithmometers, designed by a Swiss engineer named Erwin Jahnz, had gone on sale in the 1910s.[49] The STW's electric motor, too, could be traced back to pioneers such as Samuel Herzstark, who had been active at the turn of the century. (Electrical drives made calculators not only faster but also more accessible. Arnold Lowan, the administrative chief of the Mathematical Tables Project, had argued that providing electric Fridens to disabled computers would let them work more rapidly than their able-bodied counterparts with their manual machines. His request was denied.)[50]

The STW deviated from arithmometers of times past mostly by way of its imposing, 130-button keyboard. A large grid of keys, comprising columns labeled from 1 to 9, controlled the STW's inputs in lieu of the more usual sliders. A smaller keypad on the left of the machine handled multipliers and divisors, and the remaining keys sprinkled around the periphery of the machine told it what to do with all those numbers.[51] But even this seeming innovation was old

O A Friden STW-10 made around 1954. The STW-10's multitudinous keys allowed the user to enter multiple digits at the same time, rather than having to key one at a time as on a ten-key, telephone-style keypad. *"Friden Model STW 10 Calculating Machine." National Museum of American History. Washington, DC: Smithsonian Institution, 1954. https://americanhistory.si.edu/collections/ search/object/nmah_690861. Image courtesy of Division of Medicine and Science, National Museum of American History, Smithsonian Institution.*

hat. The STW's keyboard bore a striking resemblance to that of an earlier calculator, the Monroe "High-Speed Adding Calculator" of 1914, and that machine, in turn, had been substantially inspired by the "comptometer" adding machines of the late nineteenth century.[52] By the time the Friden STW came along, its keyboard layout was already more than half a century old.

JOHNSON WORKED THROUGH the mission's three planned orbits minute by minute, her Friden chattering mechanically as she did so. Taking the same equations that the IBM had used to predict the capsule's trajectory, for each sixty-second interval she computed eleven separate values to eight significant figures[*] and compared them to the computer's calculations.[53] It was a complex task undertaken with a simple tool, and the stakes were unimaginably high. An incorrect calculation could put Glenn into the wrong orbit or bring him back to Earth far from the Navy ships waiting to retrieve his capsule. Any delay or disaster at this late stage would hand victory in the space race to the USSR.[†]

With a day and a half to spare, Johnson's Friden fell silent. The

---

[*] A "significant figure" is a digit that usefully contributes to the meaning of a longer number. Leading and trailing zeroes are normally ignored, so that 01234 and 0.0001234000 are both said to have four significant figures. 1.0234 has *five* significant figures, since the zero influences the placement (and thereby the meaning) of the digits either side of it.

[†] The Pentagon gave more than a little thought to turning any disaster during Glenn's flight to its advantage. Had *Friendship 7* burned up in the atmosphere or skipped off it into space, the proposal was to point the finger at Cuba, the communist bugbear in America's backyard, and proceed to depose Fidel Castro. The plan was never officially ratified, but had it been so, Katherine Johnson's calculations would have had much more riding on them than the safety of a single astronaut.

stacks of paper on her desk attested to numbers that achieved "very good agreement," in the scientific parlance, with those of the IBM.[54] She had checked, and Glenn could go.

JOHNSON'S FRIDEN could have been replaced by any number of competing machines without affecting her work. There were the "rotary" machines, like the Friden, that aped Charles Xavier Thomas's arithmometer. There were "pin-wheel" calculators that replaced the arithmometer's stepped drums with disks from which different numbers or pins, or teeth, protruded.[55] Even the humble comptometer, whose first prototype had been installed in a macaroni box, had since matured into a respectable, full-fledged calculator.[56] All were holdovers from the nineteenth century, and all worked in much the same way.

Why, then, were calculators so conservative in their appearance and their functionality? After all, the electronic computer had shown that there was a new way to build calculating machines—one that did away with gears to be oiled and handles to be cranked, that worked near silently and with astounding speed.

Back in 1949, the Mathematical Tables Project had inadvertently accelerated the transition to this brave new world. John von Neumann, a Hungarian American scientist of prodigious intellect and an alumnus of the Manhattan Project to build the first atomic bomb, had approached the MTP with a classic problem of economics: how to feed a given number of people as cheaply as possible with a fixed set of different ingredients.[57] Twenty-five MTP computers toiled over the problem for twenty-one days, but von Neumann did not really care about their solution; instead, he wanted to confirm his suspicion that electronic computers were superior to the human variety. With the MTP's year and a half of computing time in mind, he calculated that ENIAC, the United States' first electronic computer, could have found the same answer in nine hours. Perhaps not coincidentally, the MTP was shuttered later that same year.[58]

Even in 1962, Katherine Johnson's last-minute communion with her Friden STW was less about any intangible qualities that a human might have brought to bear than it was about confirming what an electronic computer had already decided—that the third Mercury flight would go where it was supposed to go and return intact. Plainly, the electronic computer was the future, and yet the calculator languished in the past.

The issues, fundamentally, were of size and cost. ENIAC might have been more than a thousand times faster than a human computer, but it weighed thirty tons, cost the best part of half a million dollars, and occupied a room as big as a family home.[59] Fifteen years later, NASA's much-advanced IBM 7090 *still* needed a room to itself, and its price tag had ballooned to $2.9 million, or almost $26 million in today's money.[60] The gulf between calculators and computers had never been larger—and, yet, within a decade of Katherine Johnson's heroic feat of computation, it would be almost completely erased.

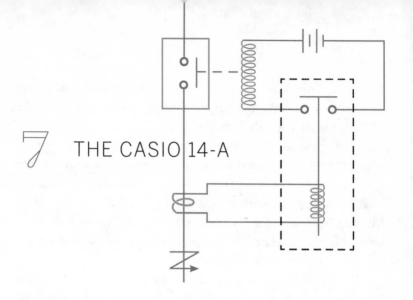

7 THE CASIO 14-A

I T WOULD NOT FIT. Late in 1957, as Tadao and Toshio Kashio waited to load their desk-sized calculator onto a plane at Tokyo's Haneda Airport, they were told they would have to dismantle it first. If you can just remove this top part here, an attendant told them, gesturing to the keyboard and display, it will fit on the airplane. If we do that, the Kashios pleaded, it may break. And so it did. When they arrived at Taiyo Sales in Sapporo to demonstrate their reassembled machine, it refused to work, and the brothers had to fall back on a slideshow.[1] The future of their calculator, and that of the Kashio family business, seemed to hang in the balance.

T HE KASHIO BROTHERS— Tadao, Toshio, Kazuo, and Yukio—were quite different from the inventors and engineers who had gone before them.[2] They did not have Blaise Pascal's mathematical prowess, or Samuel Morland's royal connections. They did not have Charles Xavier Thomas's considerable income, or Curt Herzstark's family tradition of engineering. What they did have was a finger-mounted cigarette holder that let Japan's workers get their nicotine fix both on and off the job.

The Kashios hailed from a village on Japan's southern island of Shikoku, where their parents farmed rice paddies. In 1923, however, in the aftermath of an earthquake that flattened Tokyo and Yokohama and left an estimated 140,000 dead, Shigeru Kashio moved his young family to the country's capital to help rebuild it.[3] To save money on the commute to his construction job, Shigeru walked for up to five hours each day rather than take public transport.[4]

Tadao, Shigeru's eldest son, grew up to become a metalworker and machinist. Although he was too sickly to join the military when the Second World War intruded, Tadao joined the war effort by making airplane parts on a lathe in the family's garden shed. That enterprise came to a crashing halt when the Kashio house was destroyed by American bombers. Japan's surrender, after the destruction of Hiroshima and Nagasaki, seemingly dampened any hopes of a revival. And yet, shortly after the war, Tadao was offered a milling machine located some three hundred kilometers from Tokyo, and he grabbed the chance to resuscitate the family firm. His father hitched a hand-cart to his bicycle and spent several weeks dragging the machine back home. Kashio Manufacturing was born.[5]

At first, Kashio's product line was decidedly eclectic: now joined by his younger brother Toshio, Tadao made gears, parts for microscopes, and hot plates for cooking.[6] Toshio, on the other hand, who idolized Thomas Edison and who had learned his way around electrical circuitry at Japan's Ministry of Communication, had bigger ideas. Even so, Toshio's first hit product was far removed from electrical wizardry: the so-called *yubiwa*, or "ring pipe" was nothing more than a chromed finger ring with a cigarette holder soldered to it.[7] And yet it was a runaway success. Japan's workers could smoke their unfiltered cigarettes all the way to the end while working, then enjoy a relaxing puff afterward in the *sentō*, or bathhouse, without getting their tobacco wet.[8]

The profits from the *yubiwa* pipe gave the Kashios room to breathe, and Toshio room to think. Back in 1946, he had read a newspaper article that chronicled the battle of wits between Kiyoshi Matsuzaki on his abacus and Tom Wood of the U.S. Army on his electric calculator.[9] The West had been captivated by Matsuzaki's victory, but Toshio was more intrigued by the losing machine. His interest was piqued again three years later at a business expo in Ginza, Tokyo, where Tadao and Toshio saw a clutch of similar machines in person. All had been imported, since impoverished postwar Japan lacked the industrial base to manufacture motor-driven calculators. (That would not change for another decade.)[10] Toshio decided then that Kashio Manufacturing would build an electric calculator, economy be damned, and that it would do so in a completely new way.[11]

FOUR YEARS AND ten prototypes later, Toshio and Tadao finished their calculator. Figuratively speaking, it was revolutionary; in literal terms, it was the opposite, since this was the first automatic calculator anywhere in the world that was *not* driven by rotating gears or motors but rather a kind of electromagnet called a solenoid. Familiar to Toshio from his days at the Ministry of Communication, a solenoid is a cylindrical coil of wire that, when energized by an electrical current, propels a metal "plunger" along its length.[12] The plunger, in turn, can be used to trigger an external system: to open a valve controlling the flow of a fluid, for example, or to lock a door, or to start a car. Or indeed, in Toshio's case, to trip another electrical switch. The Kashios' calculator was a symphony of solenoids and switches, wired together in series and in parallel to create circuits of ever-increasing complexity.[13] There was not a gear or a motor to be seen.

In 1955, the Kashio brothers demonstrated their calculator at

Bunshodo Corporation, a Tokyo office supply company.[14] It worked perfectly, and quietly, too, lacking the obtrusive racket generated by most electric calculators. But their potential customer had a question: why, having completed a multiplication, could one not simply multiply the result a second time, and a third, and so on? This had been a feature of mechanical calculators since Pascal's time and yet the Kashios' machine had to be reset after each arithmetical operation.[15] Reworking the calculator to satisfy Bunshodo's request cost the Kashios another year of development—and then, with the machine almost ready to go, Toshio declared that he wanted to throw it out and start over. Solenoids and switches, he declared, were too fiddly for mass production. *Relays* were the future.[16]

IN TEARING DOWN HIS solenoid calculator, Toshio Kashio was unwittingly joining a much larger movement in the design and development of computers. The roots of that movement, in turn, lay all the way back in the age of the telegraph—the first communication system that relied not on line of sight or a horse and rider to convey information but rather on the new wonder of electricity.

First, a point of etymology. The word "telegraph" means to write at a distance, and, although the term is synonymous with the dits and dahs of the electrical telegraph, humanity had been telegraphing since long before that device was ever conceived.[17] Humans have been using smoke signals since prehistoric times, for example, and flags, it turns out, are even better.[18] The use of flags to exchange messages between ships is well known, but land-based flag networks were also found in the Americas, Europe, Africa, and Asia. At one point during the early nineteenth century, France, the erstwhile leader in such things, hosted more than five hundred "optical telegraph" stations within its borders.[19]

Then came electricity. The first suggestion that a wire could carry not just a current but a message was published in the pages of *Scot Magazine* as early as 1753, but the lack of reliable power sources stymied that and many later proposals.[20] It was only in the early years of the nineteenth century, when one Samuel Morse combined a simple electrical circuit with an ingenious coding scheme, that the electrical telegraph came of age. With a switch at one end of the line and an electromagnet at the other, Morse's telegraph turned a series of on-off pulses into a dot-dashed line drawn by a pencil, which could then be decrypted by a reader familiar with Morse's eponymous code. Later, the pencil and paper tape would be replaced by a "sounder" that emitted a tapping noise.[21]

The problem that Morse and others faced was that of range. All wires dissipate current to a greater or lesser degree, so that after a certain distance the receiving mechanism will stop working. Lightbulbs will flicker and dim; telegraph sounders will stop tapping. What was needed was a way to transform a weak incoming signal into a strong outgoing one so that a series of telegraph lines could be daisy-chained together over long distances. What was needed was the *relay*.

Both Edward Davy, an English inventor, and an American named Joseph Henry, later to be the Smithsonian's first secretary, have convincing claims to the earliest working relays.[22] Their respective devices were conceptually identical too. Like Morse's telegraph, a switch at one end of a telegraph circuit controlled an electromagnet at the other. But rather than pushing a pencil onto a moving tape, or tapping out an audible rhythm, that electromagnet was in turn connected to *another* switch—a switch that operated a second telegraph circuit with its own dedicated power source. As the telegraphist tapped out a message on the first switch, the electromagnet opened and closed in sympathy, actuating the switch at the start of the second circuit, so that the dots and dashes of their message were transmitted onward, at full power, on the new circuit.[23] Nor was there any reason to stop there. Additional circuits,

O A telegraph relay made by the Western Electric Company around 1903. The black cylinders house electromagnets that, when energized, attract the vertical black plate and hence close the U-shaped brass switch above the cylinders. When the current to the cylinders is cut off, the plate is pulled away from them by a spring, and the switch reopens. *Western Electric Company. "Telegraph Relay (1997-1-0591)." Collection of Historical Scientific Instruments. Cambridge, MA: Harvard University, 1903. http://waywiser.fas.harvard.edu/objects/15028/ telegraph-relay. Image courtesy of Collection of Historical Scientific Instruments, Harvard University.*

or *relays*, could be added as often as necessary for a telegraph line to cross a county, a state, or a country. It was the final part of the telegraph puzzle.

But the relay had another use, too, one that neither Henry, Davy, nor Morse had anticipated. Consider this: sending any sufficiently strong input signal to a relay results in a closed-output circuit, while any sufficiently weak input signal results in an open circuit. If we label the closed case as "0" and the open case as "1," it becomes apparent that the humble relay can bend messy, entropic reality to the service of pristine binary logic.

O———O

"MA BELL," the hulking American Telephone and Telegraph Company, was the product of Alexander Graham Bell's quest to build an improved telegraph. The Scotsman, who, in the 1870s, had immigrated to Canada and then to the United States, had the idea that it might be possible to connect multiple telegraph senders and receivers over a single line. Instead of simply making and breaking the circuit, Bell proposed instead to send audible tones, converted into oscillating electric currents, along the wire.[24] Each telegraph key would emit a particular musical note, and a receiver tuned to that same frequency would pick up only the dots and dashes sent by the associated key.[25]

Bell's "harmonic telegraph" has been largely lost to history, but the lessons he learned while working on it led him to patent the first practical telephone system.[26] Now, Bell's instrument was neither the first working telephone nor even, possibly, the first to be patented, but he was the first to build a viable enterprise from his invention.[27] And it was in the bowels of AT&T, the resulting corporate behemoth, that the life of the relay would enter its second act.

In 1925, AT&T centralized its research staff under the banner of Bell Telephone Laboratories.[28] It was there, twelve years later, that an engineer named George Stibitz rescued a pair of relays, still used widely in telephone exchanges, from the Bell Labs scrap heap and took them home to his kitchen. There, for reasons that even Stibitz, interviewed many times about his work, seemed to have forgotten, he wired the relays together with a pair of flashlight bulbs and some batteries to make a circuit that could add one and one to make two. His wife, Dorothea, later dubbed it the "Model K," for "kitchen."[29]

Soon after, Stibitz's boss, Thornton Fry, asked him if the Model K could be made to work with complex numbers. Fry did not mean numbers bigger than two, the limit of the Model K's circuitry, but rather a very specific and lushly exotic species of number made up of separate "real" and "imaginary" components. Written as $x + iy$, where x is the real component and y is the imaginary one, the complexity of the complex number lies in the fact that i is equal to the square root of negative one—which, in standard arithmetic, has no real value.[30] Complex numbers do have applications in physics and engineering, but even simple operations such as multiplication are more time-consuming than for regular, "real" numbers.[31] Fry needed between five and ten human computers to keep up with his engineers' demands for complex-number arithmetic.[32]

Stibitz obliged, building a relay-based calculator that could add, subtract, multiply, and divide complex numbers. Completed late in 1939, the machine contained more than four hundred relays and took around a minute to multiply a pair of complex numbers—which was slow, but still many times faster than a human.[33] And as Stibitz and Fry demonstrated in September 1940 at the American Mathematical Society in New York, the "Complex Number Calculator" could also be hooked up to a telephone line and driven using a kind of electric typewriter called a teletype. It was the first ever public demonstration of remote computing.[34]

In designing the Complex Number Calculator, Stibitz may have been inspired by the work of Claude Shannon, a math student at the time, who would later marry a Bell Labs computer named Mary Elizabeth Moore.[35] Shannon's master's thesis, published in 1938, described how relays could be used to solve any problem described in terms of Boolean algebra.[36] In that statement lies a rabbit hole of considerable depth: Boolean algebra was the invention of a nineteenth-century English mathematician named George Boole, and it provided a formal way to write down and manipulate if-then statements such as "if x and y are true, then z is true" or "if a or b is true, then c is false."[37] It is, essentially, a kind of math dedicated to only two values—0, meaning false, and 1, meaning true—but, as

Gottfried Leibniz had realized two centuries earlier, those two values can be combined to represent any other kind of number one might need.[38] Here, for example, are binary counterparts of the decimal numbers from 0 to 8:

Binary	Decimal
0000	0
0001	1
0010	2
0011	3
0100	4
0101	5
0110	6
0111	7
1000	8

By combining Leibniz's binary numbers and Boole's binary algebra, it becomes possible to construct expressions that will add, subtract, multiply, or divide any kind of number at all: whole numbers; fractions; complex numbers. And if we consider that we can map numbers onto *letters*—imagine that 0 corresponds to "A," 1 corresponds to "B," 2 corresponds to "C," and so on—then we can break out from the mathematical realm and start to manipulate letters, words, and language. Shannon's insight was that relays were the perfect building blocks to bring Boolean logic into the real world. His thesis, which has been called "possibly the most important, and also the most famous, master's thesis of the century," was nothing less than a road map for building computers. Or, indeed, for building calculators.[39]

O———O

F OR A BRIEF MOMENT, running parallel to the turmoil of WWII, relays were everywhere. Bell Labs built four more relay computers during and immediately after the war.[40] And in Germany, between 1938 and 1941, a mechanical engineering student named Konrad Zuse built an enormous 2,400-relay machine called the Z3. It could store and manipulate up to sixty-four twenty-two-digit decimal numbers, but wartime shortages forced Zuse to fall back on mechanical components for its follow-up, the Z4.[41] But perhaps the most famous of all relay computers was a product of what might be called the military-industrial-academic-aesthetic complex: the Automatic Sequence Controlled Calculator of 1943 was imagined by a Harvard professor, built by IBM, sculpted by a famous designer, and put to work by the U.S. Navy.[42]

Howard Aiken, working toward a PhD in physics at Harvard University, was running out of patience with the tools available to him. The Marchants and Fridens found on many a human computer's desk could not handle the large numbers that occurred in scientific calculations, nor could they navigate the complicated algorithms needed to compute certain results. The slide rule, too, was showing its age. A growing number of scientific theorems and instruments could be adequately addressed only by "nonlinear" equations that could not be solved by multiplication and division alone.[43] To simulate the propagation of radio waves in the upper atmosphere or to tackle Einstein's theory of relativity with a slide rule was to paddle up the Mississippi with a Popsicle stick.

In a 1937 memo that invoked John Napier, Wilhelm Leibniz, Blaise Pascal, and Charles Babbage, Aiken imagined a hypothetical calculating machine that could handle numbers of dizzying precision, that would be driven by punched cards or perforated paper tape, and that would iterate happily toward solutions without human supervision.[44] Aiken made a point of comparing his "automatic calculating machine" to the tabulating machines then offered by IBM, which had to be rewired for each new step of an algorithm.[45] IBM had been around in one form or another since its founder, Herman Hollerith, had first used punched cards to help tabulate the 1890 U.S. census, but its machines had always been designed for breadth rather than depth.[46] An IBM tabulator could compute the monthly payroll for thousands of employees, but it balked at more intricate calculations.

Through a chain of contacts at Harvard, Aiken was put in touch with IBM's formidable chairman, Thomas J. Watson. Goaded, perhaps, by Aiken's unflattering appraisal of the machines that underwrote IBM's bottom line, Watson agreed to build and pay for Aiken's machine if Harvard would promise to use it only for research. IBM would learn how to build computers, and Aiken would get his machine.[47]

The completed Automatic Sequence Controlled Calculator, trucked to Harvard's physics department early in 1944, was a monster.[48] The guts of the machine held more than three thousand relays and five hundred miles of wire allied to seventy-two mechanical accumulators. It weighed almost ten thousand pounds and comprised more than three-quarters of a million parts.[49] An IBM brochure published the following year, after the "Harvard Mark I" had been unveiled to considerable fanfare, tried gamely to paint the machine as "of light weight, trim appearance," only to clarify, in the next clause, that it measured "51 feet long and 8 feet high, bearing an interlocking panel of small gears, counters, switches, and control circuits."[50] Even a sleek, steel-and-glass enclosure designed by Norman Bel Geddes, the Jony Ive of his day, could not disguise the ASCC's bulk.[51] Aiken himself was dismissive of this $50,000 sop to aesthetics, reasoning that he would rather have had another $50,000 worth of computing power instead.[52]

Almost immediately, the U.S. Navy requisitioned the ASCC and put Aiken, a Navy reservist, in charge of it.[53] The machine spent

○ The Automatic Sequence Controlled Calculator, also known as the Harvard Mark I, in 1944, prior to the installation of its Norman Bel Geddes–designed enclosure. Numbers to be operated on were input using the arrays of rotary switches to the left, while operations were synchronized by the motorized shaft running along the bottom of the machine. *International Business Machines, Harvard Computation Laboratory, and Cruft Laboratory. "IBM ASCC-Mark I Photo Album: Left-Right Panorama of Mark I during Installation (Lib.1964-056)." Collection of Historical Scientific Instruments. Cambridge, MA: Harvard University, 1944. http://waywiser.fas.harvard.edu/ objects/20119/ibm-asccmark-i-photo-album-leftright-panorama-of-mark-i-d. Image courtesy of Collection of Historical Scientific Instruments, Harvard University.*

the war computing gunnery tables and optimizing ship designs, and may even have simulated America's first atomic bomb.[54] Among the naval personnel who ran the ASCC during this time was a lieutenant named Grace Hopper, who showed both an affinity for working with the machine and a sympathy for its human wranglers. It was Hopper who collated and edited the ASCC's five-hundred-page manual, and who would, later in her career, create the first "compiler"—a software program that lets coders write applications not in the terse, error-prone language of the machine itself but rather in high-level, human-readable terms.[55] Software development as it is today owes an enormous debt of gratitude to Hopper's work.

There is a striking comparison to be drawn between Howard Aiken and Grace Hopper on the one hand and Charles Babbage and Ada Lovelace on the other. One of Aiken's many inspirations, Babbage was the irascible nineteenth-century genius behind the fabled "analytical engine," a mechanical computer that, had it ever been completed, would have pre-dated all other general-purpose computers, the ASCC included.[56] Ada King, countess of Lovelace and the only child of the infamously "mad, bad, and dangerous to know" Lord Byron, saw more clearly than Babbage himself how the analytical engine could be turned to myriad different purposes.[57] Aiken and Babbage may have invented the hardware, but Hopper and Lovelace understood that software was the key to making the best of it.

ON THE OTHER SIDE of the Pacific, Japan, too, had fallen under the relay's spell. Since 1934, an electrical engineer named Akira Nakashima had been trying to turn the trial-and-error art of circuit design into a formal mathematical process. The electrical properties of the telephone networks he encountered at his job at Nippon Electric Company, or NEC, had been studied to within an inch of their lives, but the complex relay mechanisms that connected them were largely ignored, and consigned to mechanics rather than scientists. By publishing his theory of relay circuits in 1936, Nakashima came agonizingly close to making the same connection between relays and Boolean logic that Claude Shannon would describe just two years later, but the Japanese engineer failed to see the link.[58] Even so, Nakashima had fired the starting gun on Japan's own relay race.

The country's first relay computer was built at the Ministry of

Communication's Electrotechnical Laboratory. The ETL Mark I of 1952 was a modest machine, but it was promising enough for a Mark II, comprising a staggering 22,000 relays, to follow in 1955.[59] That computer, in turn, was faster than the "Harvard Mark II" that had followed Howard Aiken's ASCC, the Harvard Mark I, in 1947.[60] Japan was behind, but it was catching up. And so, a handful of years later, when Toshio Kashio was looking for a replacement for clunky, finicky solenoids, the relay was waiting to greet him.

The Kashios' rebuilt calculator was driven by 342 relays, much reduced in size from the earliest telegraphic models, that tapped away like proverbial Shakespearean monkeys as they ground through a multiplication or a division. Relays aside, the Casio 14-A had no moving parts. Where other calculators used odometer-like mechanical displays to show their input and output values, the 14-A boasted a grid of numerical digits that were lit up or extinguished to represent

O The Casio 14-A. No gears, no shafts, no motors; this was the first calculator to compute answers using relays rather than mechanical components. © *Casio Computer Co., Ltd.*

both the current input number and the result of any calculation. In addition, the 14-A was one of the earliest calculators to deviate from the serried columns of a full-scale comptometer keyboard, in that it employed a simple ten-key pad labeled from 0 to 9.[61] The 14-A's only real foible was its size. It was a desk calculator not in the sense that it fitted on one, but rather that it *was* a desk that happened to weigh 140 kilos and that consumed three hundred watts of power.[62]

Nevertheless, the Kashios' machine surprised the Americans— and the Europeans, and everybody else. It was the world's first purely electrical calculator, with nary a gear or a cam to its name. When it worked, it was a marvel; and, when the Kashios tried to demonstrate it in Sapporo, it did not.

But broken or not, the 14-A had made an impression. Shortly after returning from Sapporo, the Kashios were approached by a representative from Taiyo's parent company, Uchida Yoko, to offer them a distribution deal. Taiyo specialized in mechanical calculators, but Uchida Yoko saw the potential in the Kashios' hulking gray-green device. Accordingly, in June 1957, Kashio Manufacturing became the Casio Computer Company, and the 14-A went on sale.[63] The change in spelling is mysterious: some sources say the brothers anglicized their name so that it would scan better in English; another says that it was a reference to the constellation of Cassiopeia; and yet another that the new spelling simply "looked cooler."[64]

Whatever the reason, the Casio 14-A was a success. It was not cheap—at ¥485,000, it would have cost $1,347 then or $12,900 today—but it established a reputation as a quiet, reliable machine and found its way into government, financial, and general commercial use.[65] Toshio added a square-root function in 1959, necessitating the addition of eight relays to the calculator and ¥165,000 to its price.[66] By 1962, Casio Computer was making a healthy $1.7 million in revenue each year, selling both calculators and a new line of programmable, relay-based computers.[67]

ALTHOUGH RELAY CALCULATORS WERE quieter and faster than arithmometers, comptometers, and the like, they were still mechanical devices at heart. Each button pressed required hundreds of relays to operate in perfect harmony, their switches free of obstruction and their contacts clean. A single relay clogged by dust or dirt could lead to an erroneous result—and worse, if the calculator were to be knocked or jostled sometime later, and that speck of dirt dislodged, the same operation would then work perfectly with no indication that anything had ever gone wrong.[68] Not a problem for a telephone exchange, where a dropped call was merely an inconvenience, but a bug in a gunnery table or an orbital trajectory could be deadly.[69]

It was for this reason that Bell Labs' relay computers were saddled with complex error-checking circuits. A study done there found that relays tended to fail once every 2 to 3 million cycles, which equated to four or five errors per day in Bell's more advanced computers.[70] Howard Aiken's Mark I and Mark II computers had no such safeguards, which led to one of the most celebrated incidents in computing history.[71] While running an error-checking calculation on the Mark II, an unnamed engineer* tracked down a persistent failure to a single relay—relay #70 in Panel F, it turned out—which had been incapacitated by an errant moth.[72] Grace Hopper taped the insect into her logbook and recorded it as "First actual case of bug being found."[73] The concept of a computer "bug," already in common currency, was immortalized.

The relay's shortcomings meant that its reign in the world of

---

* The incident is recounted in a 1969 interview conducted with Grace Hopper. The engineer's name is redacted from the interview transcript.

calculation would be short. Better switches were on the way, to be embraced by the makers of computers and calculators alike, but the Kashios had taken their eyes off the ball. Or rather, they had their eyes on a different ball altogether. Having installed themselves at the head of Japan's calculator pecking order, the four brothers were increasingly partial to the golf course rather than the factory.[74] The future of the calculator would lie in other hands.

THE SUMLOCK
ANITA

HE SAME YEAR that Katherine Johnson double-checked John Glenn's flight plan on her Friden STW-10, a British company launched a calculator that made the Friden look positively antiquated. That calculator, the Sumlock ANITA, also made the Friden *sound* antiquated: the British model was the world's first electronic calculator, with no moving parts beyond its keyboard, and it was completely silent. The boxy, bulky ANITA was the calculator's first step into a solid-state future.

FOUNDED IN 1878 to make ticket punches for tram and bus conductors, the Sumlock Company* of postwar Britain was a paro-

---

* Sumlock's original incarnation was called Bell Punch. Its history of mergers, acquisitions, and spin-offs is complex enough to fill a book of its own, so in the interests of brevity, "Sumlock" will be used here to mean any part of the corporate family tree that took an interest in designing, building, or marketing the ANITA.

chial echo of the corporate giants on the other side of the Atlantic.[1] Where IBM built gleaming mainframes, Sumlock peddled comptometers first developed in the nineteenth century.[2] Where AT&T's Bell Labs inhabited a neoclassical tower in Manhattan, Sumlock's engineers toiled in a converted air-raid shelter in London's suburban hinterlands.[3] And where Burroughs, NCR, and General Electric worked with the Federal Reserve to develop a nationwide system of machine-readable checks, Sumlock fretted about competition from a chain of tea shops.[4]

On the flip side, the British firm was nothing if not adaptable. Sumlock had rebounded from a fire that leveled its factory at the end of the nineteenth century to become Britain's premier manufacturer of taximeters and racecourse betting boards, or "totalizators." It acquired a paper mill to supply the tram and bus companies that used its punches, and became the world's largest printer of tickets. During World War II, it successfully pivoted to the production of mechanical aiming systems for Britain's bombers and warships.[5] And then there were the Sumlock calculators.

In 1936, Sumlock had bought the rights to a compact, comptometer-style adding machine called the "Petometer." The renamed "Plus Adder" went on sale the next year, to be followed in 1940 by a full-sized adding machine. Later still, Sumlock acquired the British arm of the Comptometer Corporation itself, and by 1961, the original Chicagoan company had shut down its factories in favor of simply importing comptometers made by Sumlock in the UK.[6] Sumlock was riding high, but something new and threatening loomed on the horizon.

In the aftermath of the Second World War, the mathematicians and engineers of Bletchley Park had scattered to the far corners of the U.K.'s academic establishment. At the universities of Manchester and Cambridge and at the National Physical Laboratory and Birkbeck College in London, Alan Turing and others were busy planting the seeds of Britain's nascent computing industry.[7] It was a source of some national pride that, by 1950, three of those four groups had succeeded in building a functional electronic computer while a dozen similar projects in the United States still struggled to catch up. In the whole of the United States at that time there existed only a single working electronic computer in the form of the ENIAC.[8]

The British private sector was equally precocious. As early as 1947, Lyons, a catering company known as much for its forward-looking management as for its ubiquitous, stylish tea shops, had sent two employees to the States to learn more about the "electronic brains" then in development for the U.S. military. The Lyons pair were denied access to all these American contraptions, but they somehow learned about Britain's efforts to build similar machines with public money. On their return, the Lyons board was convinced to part with a cool hundred thousand pounds to fund a computer project, and in 1951, the company unveiled an electronic computer called the Lyons Electronic Office.[9]

Belatedly, Sumlock's management realized that their line of mechanical calculators was at risk: if a computer could turn around Lyons' entire payroll in seconds, what use was there for an adding machine? Dashing to make up for lost time, in 1956 Sumlock set up an electronics department and cast about for an engineer to run it.[10] That engineer was to be one Norbert Kitz, a veteran of the computer projects at both Birkbeck College and the National Physical Laboratory, and he had grand plans to build a calculator like none other. Kitz knew his stuff: he had constructed the first iteration of Birkbeck's computer, and at NPL he had worked on a machine called Pilot ACE (for "automatic computing engine") that was, in turn, modeled on a proposal by the brilliant, doomed Alan Turing.[11] As such, Kitz knew that the relay was not long for the computing world, and he had already identified a perfect replacement for it. What he needed—what Sumlock needed—was the vacuum tube.

○——○

THOMAS ALVA EDISON (you can call him Al, since his family did too) was, like his competitor Alexander Graham Bell, an immigrant to the USA. (Unlike Bell, Edison was an ancestral American: two generations earlier, Edison's grandfather had fled to Canada during the American Revolution.)[12] Between them, Edison and Bell wired America for light and for sound, and whereas the relay had arisen in Bell's fief of the telegraph and telephone, it would be Edison's electric empire that gave birth to the vacuum tube.[13]

In 1883, Edison was basking in the glow of success. His system of steam-driven dynamos and incandescent lightbulbs had proved to be as capable of illuminating cities, such as New York and Chicago, as they were a steamship's cabins, such as those of the SS *Columbia* with its four Edison dynamos and 120 electric lights.[14] But the prolific "Wizard of Menlo Park," who would eventually amass more than a thousand patents, was unhappy with the performance of his bamboo-filament lightbulbs.[15] Too often, a bulb would end up blackened and opaque, with only a peculiar sliver of clear glass remaining. Edison wanted to know why.[16]

On further inspection, it seemed to Edison that the bulb's negative terminal, its "cathode," to which one end of the filament was connected, was emitting some kind of material in all directions that then blackened the inside of the glass. Only the bulb's "anode," or positive terminal, at the other end of the filament, stood in the way of this stream of material and protected the glass behind it.

To investigate further, Edison made a series of new bulbs, each one with a metal plate positioned between the filament's terminals (a typical arrangement is shown in Figure 32, below), to see what effect the plates might have on the invisible material being emitted by the negative terminal. And, largely because he could, he connected each bulb's plate to a galvanometer—a device to measure electrical current—to see if anything interesting would happen.[17] He was rewarded.

In a conventional bulb, Edison knew that current flowed from

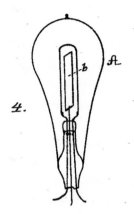

○ An Edison bulb with a plate, *b*, nestled inside the filament. This image is taken from U.S. patent 307,031, in which Edison described the flow of current across the empty space within the bulb and proposed to use the phenomenon to influence an attached electrical circuit. *Edison, Thomas A. Electrical Indicator. 307,031 A. USA, issued October 1884. http://www.google.com/patents/ US307031A.*

one terminal to the other, heating the filament as it did so and causing it to glow as a result. In his modified bulbs, Edison discovered a new phenomenon: when he connected a galvanometer between a bulb's plate and its positive terminal, the galvanometer showed that a current was passing through it. But how? The plate had no physical connection to the negative terminal, and so there should have been no way for the current to pass. Somehow, current was traveling across the empty space between the negative terminal and the plate and causing the galvanometer's needle to move in sympathy. Curiouser still, when Edison moved the galvanometer from the positive terminal to the negative terminal, it lay dormant: his mysterious current did *not* flow from the positive terminal to the plate. Two more observations followed: the current through the plate varied in concert with the current through the bulb; and the entire effect seemed to be amplified as the bulb grew hot with use.

Later, scientists would realize that this free-space current was caused by charged particles first called "cathode rays," then "corpuscles," and finally "electrons," that leapt off the hot cathode and onto the plate.[18] They dubbed the phenomenon "thermionic emission," and figured out that these free electrons had also been the cause of

Edison's blackened bulbs. But those collective revelations were still a decade or more away, so Edison, never a man to let a discovery go unpatented, suggested that a bulb, plate, and galvanometer could be used to measure and perhaps control the current flowing through an electrical system, but he did not pursue the project.[19] Distracted by the ongoing rollout of his lighting system, he filed his patent and went on his way.[20]

The "Edison effect" did not go unnoticed. First, in 1904, an Englishman named John Ambrose Fleming patented the use of an Edison bulb to convert an alternating current, which oscillates between positive and negative flow, to a direct current that flows in one direction only. Recalling that Edison's current flowed only from the negative terminal to the plate, but not from the positive terminal to the plate, Fleming connected the bulb's terminals to an alternating input signal and its plate to an output circuit. When the alternating input current flowed from the negative terminal to the positive terminal, the plate picked up the incoming signal. But when the input current reversed direction to flow from positive to negative, the plate went dead. The alternating input current had been "rectified" into a direct-output current. Fleming called his invention an oscillation valve and put it to use in AM radio receivers. Today, we call it and its descendants diodes, and they are fundamental building blocks of electronic devices.[21]

A few years after Fleming's patent came two more innovations that would shape the world of electronics for the next half century. Or rather, one innovation made in two places: the archetypal vacuum tube was independently invented in Austria by Robert von Lieben and in the USA by Lee de Forest. Von Lieben, toiling in his Viennese workshop, sought to build a "telephone relay"—a device that would amplify not the staccato binary signal of a telegraph key but rather the continuously varying waveform of the human voice.[22] De Forest, in New York City, wanted to improve the behavior of Fleming's diode for use in radio receivers.[23] Both men realized that a bulb like Edison's, where current flowed from a hot cathode to a plate beside it, was a stepping-stone to a much more useful device.

That new device, later named the "triode" by analogy with "diode," worked by regulating the flow of negative charge from a hot cathode to a plate in the middle of a bulb. To achieve that control, an extra wire,

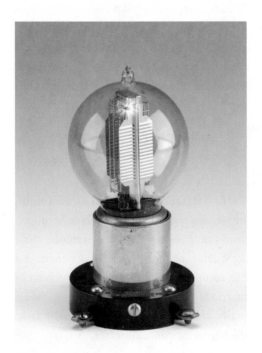

O A triode manufactured after 1922 by the General Radio Company of Cambridge, Massachusetts. In this elaborate model, the negative terminal, or cathode, lies at the center of the bulb and is sandwiched between two control grids. Beyond those lie two positive terminals, or anodes, in the form of flat plates. *General Radio Company. "Double Plate Triode (VT1150)." Collection of Historical Scientific Instruments. Cambridge, MA: Harvard University, 1922. http:// waywiser.fas.harvard.edu/objects/5142/double-plate-triode. Image courtesy of Collection of Historical Scientific Instruments, Harvard University.*

bent into a back-and-forth grid shape, was placed between the cathode and the plate. When the grid was given a small negative charge, the electrons streaming off the cathode were repelled by the grid,[*] and the current that reached the plate was reduced proportionately.[24] The grid, therefore, acted as a kind of volume control, so that small changes in the grid's charge resulted in much larger changes to the plate's output current. It was perfect for amplifying a weak telephone signal or radio broadcast.[25] So much so that, by the middle of the twentieth century, there would be more vacuum tubes than people in the USA. Telephone networks, once confined to local areas, now spanned the continental United States, and powerful radio transmitters reached across the globe.[26]

WITH THE TRIODE established as an amplifier par excellence, there was room to consider how it might be turned to other uses. From one point of view, for example, the triode looked very much like a relay, or, more broadly, a switch. A relay used an input current to complete a circuit through which an output current could pass; the triode used an input signal to regulate an output signal. Conceptually, there was little to separate the two, but in practice they could hardly be more different. A triode could turn itself on and off tens or even hundreds of thousands of times per second, while relays struggled to crack a hundred and did not last long at

---

* Von Lieben actually proposed two distinct types of tube. In one, he relied on an extra filament to control the flow of electrons from the cathode to the plate, as de Forest would do a year later. In the other, he used an electromagnet mounted outside the bulb to deflect the electrons toward or away from the plate. It was the same principle that would later underpin the cathode ray tubes, or CRTs, that were once ubiquitous in televisions and desktop computers.

such a rate.[27] Relays were power hungry and bulky; tubes were compact and parsimonious.[28] If the vacuum tube gave anything away to the relay, it was that it was a more delicate device: bulbs shattered, wires broke, and components buckled under the heat and current they were forced to endure.[29]

Even so, when an ailing tube eventually broke it could be relied upon to break in a helpful way. As Grace Hopper, Howard Aiken, and other computing pioneers had discovered, a faulty relay was worse than a broken one: a half-dead insect might free itself from a relay's magnetic embrace, or the vibrations of a passing truck might shake loose a piece of interfering dirt, so that a logical error could vanish like a ghost from the machine.[30] But triodes, like plain old incandescent lightbulbs, failed hard. A burned-out tube was broken, not misleading.[31] Fast switching, compact, and with a reliable failure mode, the vacuum tube was a superior replacement for the relay.

Gradually, the vacuum tube broadened its repertoire. In 1919, for example, a pair of British physicists named William Eccles and Frank Jordan built a tube-powered gadget that could store a single binary value—a zero or a one—and flip to the other value on demand. It was, in essence, the first electronic memory circuit.[32] Then, in 1930, one C. E. Wynn-Williams exploited the triode's ability to rapidly change states to make a cosmic ray detector. And soon after, physicists and engineers used triodes to generate the strong, clean pulses of current needed to drive radar systems: a powerful current streaming between the triode's cathode and plate was regulated by a weaker on/off current through its control grid.[33] Although radar was essentially an analog technology, there was more than a hint of the binary about the triode's role in it.

Finally, it would be one John Vincent Atanasoff, a Hungarian American professor at Iowa State College, who would first put triodes to work in a computer. In the late 1920s, as Atanasoff worked on a doctorate in physics, he had spent weeks grinding through equations that described the structure of helium atoms. With only a mechanical desktop calculator to help him crunch the numbers, the

experience instilled him with an abiding aversion to human computing. One evening early in 1937, after a frustrating day in his laboratory at Iowa State, Atanasoff took off on a drive to clear his head. He
had for years been contemplating how one might build a machine
to solve such equations, but his plans had gone nowhere. After two
hundred miles of aimless wandering, Atanasoff landed at a roadside
tavern just a little south of Moline in Illinois and, having ordered a
drink, his ideas began to coalesce.[34]

Atanasoff would recall the evening many years later. "Now, I don't
know why my mind worked then when it had not worked previously,"
he said, "but things seemed to be good and cool and quiet." As he
relaxed in the tavern, he decided that his machine would store numbers
digitally. That is, rather than use a physical quantity such as a position
on a scale or the rotation of a shaft, for the sake of precision each number
would be held as a set of individual digits. Separately, with his ground-
ing in physics rather than engineering, Atanasoff thought it would be
prudent to build his machine from electrical rather than mechanical
components.[35] This was not an easy decision: special-purpose "analog
computers," such as those made by Sumlock during the Second World
War, could already solve certain equations using rollers, gears, and pul-
leys, and Atanasoff had thought seriously about doing the same.[36]

Atanasoff ordered another drink. By the end of the night, when
he walked back to his car in the cold of the Illinois winter, he had
roughed out a circuit that could store binary numbers, with each 0 or
1 represented by an electrical charge in a component called a capacitor,* and which would send those bits to a separate processing unit
made up of triodes.[37]

---

* Unlike Eccles and Jordan's triode-based "flip-flop," which requires a constant
power supply, a capacitor will eventually lose its charge. One of Atanasoff's innovations was to periodically refresh the charge held in each capacitor so that the
contents of his computer's memory would not degrade.

Arguably, very little of this was genuinely new. The gear trains of mechanical calculators, from the time of Schickard onward, already embodied the digital representation of numbers. The arithmometer separated logic and storage by way of its stepped drums and movable carriage. Eccles and Jordan had built a binary memory circuit almost two decades earlier, and George Stibitz, on his kitchen table, had demonstrated how to make a binary adder using relays. But in choosing to make use of vacuum tubes—and, most important, by actually building his machine, with the help of a graduate student named Clifford Berry—Atanasoff became the father of electronic computing.[38] Every PC ever booted up, every calculator ever pulled out for math class, owes its existence to a thirsty professor who crossed the Iowa–Illinois state line one night in the winter of 1937.

I T IS POSSIBLE TO draw a line, if not exactly an arrow-straight one, from the Atanasoff–Berry Computer to Norbert Kitz's office. Before the war took Atanasoff away from the ABC, he had talked over its design with a physicist named John William Mauchly.[39] Mauchly's interests lay in the modern science of meteorology, established two decades earlier when an Englishman named Lewis Fry Richardson had first applied numerical algorithms to weather forecasting. Richardson, whose ideas raced ahead of his ability to execute them, had imagined calculating the weather for the entire planet with a staff of 64,000 human computers.[40] Mauchly wanted to use just one computer—a machine, this time—and was much taken by Atanasoff's ideas. Partnering with J. Presper Eckert, once Mauchly's electronics instructor, the pair set out to design what would eventually become the ENIAC, or Electronic Numerical Integrator and Computer.[41]

Built in secret for the U.S. Army, the ENIAC was declassified in 1946 to great fanfare. A front-page article in the *New York Times* hailed it as "an amazing machine which applies electronic speeds for the first time to mathematical tasks hitherto too difficult and cumbersome for solutions." Not only that, but the ENIAC could be programmed to make decisions based on the data fed into it on punched cards: "It can, for instance, compare two numbers and, depending on which one is larger, choose one of two possible courses."[42] Atanasoff's ABC, designed only to solve linear equations, had been a specialist tool. By contrast, Eckert and Mauchly's ENIAC was the first general-purpose programmable electronic computer, and neither its sponsors nor its designers were about to let anyone forget it.

Much later, two things would tarnish the ENIAC's shine. Beginning in the late 1960s, a series of patent lawsuits saw Eckert and Mauchly stripped of their mantle as inventors of the electronic computer in favor of John Atanasoff. Then, in 1975, the British government began to reveal details of the Colossus, a vacuum tube computer constructed at Bletchley Park in 1943, that could also be programmed to a limited degree.[43] In its pomp, nevertheless, the ENIAC held an absolute fascination for those who learned of it—a fascination that reached all the way to the Sumlock offices in Uxbridge, London. Even before the ENIAC had become public knowledge, J. R. Womersley, once Norbert Kitz's boss at the NPL, had made a pilgrimage to America to learn more about it. Kitz's supervisor at Birkbeck College, Andrew Booth, did the same in 1947.[44]

Accordingly, when Kitz took the job at Sumlock, his head was filled with lofty ideas of electronics and computation. Sumlock may not have wanted an actual computer, but that did not mean Kitz could not give them the most advanced calculator in the world.

○——○

N APPEARANCE, Kitz's ANITA was a curious hybrid of cash register and comptometer, as filtered through the aesthetic sensibilities of a *Star Trek* prop designer. (The calculator's name, incidentally, stood for either "A New Inspiration To Accounting" or "A New Inspiration To Arithmetic," depending on the copywriter on duty. For the remainder of this chapter, to avoid bellowing repeatedly in ALL CAPS, it will be called the Anita.) It was shaped like a large, flattened typewriter, about fifteen inches wide and eighteen deep, and aped the user interface of traditional calculators, with a large keypad for setting input numbers and a smaller one for multipliers. Rather than the movable carriage of a typewriter or

○ The Sumlock ANITA Mark VIII. (Sumlock made two versions, the VII and VIII, that differed only in minor details.) The Anita's keyboard layout mimicked that of the venerable mechanical Comptometer. *"Anita Mark VIII Desktop Electronic Calculator with Manuals." National Museum of American History. Washington, DC: Smithsonian Institution, 1961. https://americanhistory.si.edu/collections/ search/object/nmah_557378. Image courtesy of Division of Medicine and Science, National Museum of American History, Smithsonian Institution.*

arithmometer, the Anita was topped by a row of tiny neon bulbs called "Nixie" tubes, each one of which contained ten bent-wire electrodes representing the digits 0 to 9.[45] Its thirty-pound bulk confined the Anita to the desktop, but its innovative plastic case shaved some weight and protected its user from electric shocks.[46]

Inside, the Anita fizzed with novelty. It was packed with 177 "cold-cathode trigger tubes," relatives of the triode, that winked on and off as they handled the currents of its logic circuits. Also present was a single "Dekatron," an elaborate vacuum tube that activated each of its ten terminals in sequence, which acted as a central clock to regulate the Anita's internal operations. It was a clock that ticked at three thousand cycles per second, so that the Anita could add a pair of numbers in as little as thirty milliseconds.[47] A contemporary review published in Britain's *Office* magazine said that addition, subtraction, and multiplication occurred "virtually instantaneously,"* with only division letting the side down: to compute 1 ÷ 7 to twelve digits, that being the capacity of the Anita's display, took five seconds.[48] A little disappointing, perhaps, but still much faster than an electrically driven mechanical calculator. And of course, all this happened in complete silence. The metallic racket of a traditional calculator was replaced by the darting orange fireflies of the trigger tubes, lamentably hidden from view by the Anita's plastic shell.[49]

SUMLOCK LAUNCHED the Anita in 1961 at London's Business Efficiency Exhibition, with a cryptic advertisement that

---

* *Office*'s review is peppered with references to the calculator's operator as "she" and "her." Today, that might be a sign of a writer with an awareness of gender equality; in 1961, however, it was a reminder that menial office roles, such as typist or computer, were almost always filled by women.

invited the reader to visit Stand 108, "where you will find something of outstanding interest." The initial reception at home in the U.K. and abroad in Europe was generally positive. The Anita cost £355, or about a thousand dollars, which was close to the price tags of competing mechanical calculators, and its comptometer-style keyboard and operation let users adapt to it with only a minimum of training.[50] If anything, the Anita's silence presented more of a challenge than its user interface. As John Sparkes, an engineer who helped develop the calculator, recalled:

> Experienced calculator operators were slow to appreciate the technological breakthrough because they could not get used to the lack of "feel" and noise that was inherent in mechanical machines. The same phenomenon was experienced by typists changing from manual to electric typewriters.[51]

Luddites aside, the Anita's only real obstacle at the time of its launch was the currency of its home country. Britain's coinage, inherited from ancient Rome by way of the Holy Roman Empire, divided each pound into twenty shillings and each shilling into twelve pence.[52] Customary units such as the guinea, worth twenty-one shillings, complicated matters yet further.[53] A debate surrounding decimalization had been rumbling on for close to two centuries: in 1847, for example, a politician named John Bowring had urged the House of Commons to adopt the decimal counting system used by indigenous peoples and industrialized nations alike. "Every man who looked at his ten fingers," said Bowring, "saw an argument for its use, and an evidence of its practicability."[54]

British accountants were accustomed to mechanical calculators, including some of Sumlock's own, that made allowances for this archaic system, but the Anita made no such concessions. In common with most general-purpose calculators, it was a decimal machine through and through.[55] As a consequence, sales were slower at first in the Anita's home country than they were in decimalized main-land Europe.[56] It would not be until February 15, 1971, or "Decimal Day," that Britain would finally shuck off its antiquated coinage, by which time Sumlock's decimal Anita was starting to look prescient rather than problematic. The following year, *New Scientist* reported, Sumlock was comfortably the U.K.'s largest calculator manufacturer, claiming more than a third of the domestic market and far outstripping its foreign rivals.[57]

The Anita made the world sit up and take notice. By 1964, just three years after its launch, the Anita was faced with an array of electronic competitors: from America came the Friden EC-130; from Japan, the Sharp Compet CS-10A; and from Italy, the IME 84.[58] None used vacuum tubes for computation. In fact, no mass-produced calculator would ever do so again.[59] Buried within the Anita's power supply system was a single instance of a new kind of electronic component that promised to eradicate the vacuum tube in short order.[60] At a conference in 1964, Norbert Kitz, the Anita's designer, concluded his paper with a prediction, or perhaps a warning:

> Unless more development and, indeed, fundamental research, is carried out into the field of [vacuum tube] applications, the users of these components will have few, if any weapons with which to fight the onslaught of the healthy dollar fed baby known as the transistor.[61]

The future was transistorized.

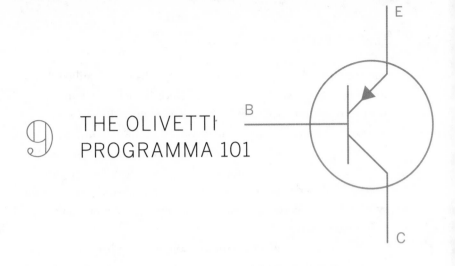

9 THE OLIVETTI PROGRAMMA 101

XT. ITALIAN VILLA, 1965. A commercial opens with a woman in a bathing suit, fresh from the pool, sitting down at a table beside a man clad in business attire. On the table is what looks like a cross between a cash register and a typewriter—except that it is *sexy*. It is organic and mechanical at the same time, with gill-like louvres, curved edges, and keys that swell up to meet the fingers like bones beneath the skin. It is the Olivetti Programma 101, the most exciting calculator in the world, and it is an imposter.

The story of the P101 begins three years earlier in Ivrea, Italy, the home of the Olivetti company. But the story of Olivetti itself, and what it hoped to gain by building the world's most advanced calculator (or the world's cheapest computer, depending on one's perspective) starts much earlier than that.

○——○

ALMOST BY ACCIDENT, postwar Italy found itself with one of the strongest economies in the world.[1] Its star performer was the Olivetti company, founded in 1908 by one Camillo Olivetti to build typewriters but which had since branched out into office furniture, adding machines, and even teletypes.[2] By 1959, Olivetti was one of the largest such enterprises in the world, and under the leadership of Camillo's son, Adriano, it was poised to grow even bigger. Adriano was set to complete the most expensive foreign takeover of an American company ever staged, in which Olivetti would swallow up Underwood, one of America's most prominent typewriter manufacturers.[3]

For Adriano, it was the culmination of a long-held ambition. In 1925, Adriano had helped a prominent leftist to escape an increasingly fascist Italy, and had found it prudent to make himself scarce shortly thereafter. He bolted to the U.S., where he would spend two years in exile.[4] He passed the time touring American factories and soaking up the techniques of mass production that kept them ahead of the rest of the world.[5] Ford Motor Company, for example, welcomed the Italian to Highland Park, Michigan, with open arms; Underwood, in Connecticut, showed him the door, sparking in Olivetti a lifelong desire to get even.[6]

Back in Benito Mussolini's Italy, Adriano was obliged to walk a tightrope between public appearances and private actions. He made a show of joining the Fascist Party, but he and his wife, Paola Levi, worked in secret to remove it from government.[7] The Second World War added considerable urgency to their cause. Mirroring the Herzstarks' predicament in Austria, Olivetti's typewriter assembly lines were now compelled to make guns for Mussolini. But Adriano was not cowed. He kept the factory's cafeterias open for those in need and arranged fake documents for those in danger.[8] He lobbied diplomats, soldiers, and even Italian royalty to raise support for Mussolini's removal. And he met, on a trip to the spy-infested capital of Switzerland, an agent for the Office of Strategic Services named Allen Dulles, who would later become director of the Central Intelligence Agency. The OSS gave Olivetti a code name: Source 660.[9]

Unfortunately for the Italian, Allen Dulles was more interested in absorbing Olivetti's local knowledge than he was in supporting his plan to oust Mussolini. But when Mussolini was abruptly sacked by King Victor Emmanuel III in the wake of Allied advances, what should have been a moment for celebration instead put Adriano in danger once again: Italy's new prime minister, one Pietro Badoglio, knew that Olivetti had championed a rival politician for the position, and promptly threw the Ivrean in jail.[10] On his release, a chastened Olivetti fled to Switzerland for the remainder of the war.[11]

BY THE 1950S, Olivetti had become one of Italy's most forward-looking companies. Adriano's distrust of fascism had coalesced into a political philosophy he called Comunità (Community), in which he imagined a patchwork of self-governing communities arranged into loose confederations.[12] His foray into politics was not a roaring success—in the 1958 elections, only a single Comunità delegate, Adriano himself, was elected to the Italian Parliament—but in Ivrea, Adriano had a ready-made laboratory in which to refine his ideas.[13]

Feeding into Adriano's philosophy was an abiding memory of his first taste of factory work. Many years before, Camillo had sent his thirteen-year-old son to work in the family factory, and Olivetti fils could not get out fast enough. He detested the hard, monotonous work of the assembly line, but came away inspired to fix it rather than avoid it.[14] Under Adriano, the company became renowned for its kindergartens, summer camps, and generous pensions, while the Ivrean campus acquired a thirty-thousand-volume library and hosted lunchtime talks from distinguished visitors.[15] So profound was Adriano's vision that his utopian factory town was placed on UNESCO's World Heritage List in 2018.[16]

But Ivrea was as much a physical experiment as a social one.

Still scarred by his apprenticeship in his father's dingy workshop, Adriano commissioned light and airy factories, offices, and homes for his workers, all styled in the glass, concrete, and steel of midcentury Modernism.[17] And like his father, who had obsessed over the "feel" of his typewriter keyboards, Adriano's enthusiasm for pitch-perfect design extended to all of Olivetti's products.[18] Before, during, and after the war, the company went on a hiring spree that saw it snap up architects, designers, poets, novelists, and painters.[19] (Milton Glaser, the graphic designer behind the "I♥NY" logo, was a typical Olivetti hire.)[20] Between them, this cadre of creatives built the hall-mark "Olivetti style"—a minimal, modern aesthetic that produced some of the most striking products and advertisements of the twentieth century. Through it all, Adriano himself had the final word* on everything.[21]

This relentless attention to detail paid off. The catalogue for a 1952 exhibition at New York's Museum of Modern Art opined that Olivetti had become "the leading corporation in the Western world in the field of design."[22] Olivetti's spry, duck-egg-blue Lettera 22 portable typewriter was a hit with writers such as Thomas Pynchon and Sylvia Plath.[23] (Bob Dylan preferred its larger sibling, the streamlined Lexikon 80.)[24] The company's line of ten-key mechanical calculators shared the good looks of its typewriters and may have pioneered the modern convention for calculator input: type a number, then an operator such as "+" or "−", and then another number to complete the operation.[25] Rather than adapt to the quirks of an arithmometer or a comptometer, users could now key in calculations exactly as they were written on the page. Olivetti, and its boss, could do no wrong.

---

* With his hands-on approach and insistence on a consistent company style, it would be easy to call Adriano Olivetti the Steve Jobs of midcentury Italy. In fact, the comparison should be made the other way round: awakened to Italian design at a conference in 1981, Jobs was strongly influenced by Marco Bellini, Ettore Sottsass, and other Olivetti designers.

And then Underwood came knocking. At first, its representatives asked to sell Olivetti's machines in the USA; and then, bashfully, they admitted that the Underwood business itself was for sale.[26] "Underwood to me was like Mecca to the Arabs," Adriano told a confidant, but the timing was not good. Olivetti was starting to feel the financial strain of its boss's magnanimity toward his workers, and that same boss now found his attention divided between his company and the Italian Parliament.[27] Nor was Underwood the choicest of prizes, saddled as it was with outdated factories, moribund products, and complacent management.[28] Even so, Adriano managed to convince Olivetti's board that the deal was worth the risk: let us buy Underwood, he said, and we will replace their old products with our new ones and use their sales force to sell them to America.[29] The deal was set in motion, with Olivetti acquiring a third of Underwood's stock in the fall of 1959, but Adriano would not live to see it completed.[30] On February 27, 1960, aboard a train to Switzerland, Olivetti's totemic chief died suddenly of a heart attack. He was fifty-eight years old.[31]

F, IN THE MONTHS BEFORE his death, Adriano had been distracted by unfinished business from the past, it was fortunate indeed that he had already laid the foundations of Olivetti's future. Back in 1954, Adriano had coaxed a brilliant young engineer named Mario Tchou to join an electronics laboratory that Olivetti was setting up in collaboration with the University of Pisa. (A lab at Ivrea, where mechanical engineers ruled the roost, was never a possibility.)[32] Four years later, Tchou and his group unveiled the Olivetti Elea 9003, a room-filling mainframe built into sleek turquoise cabinets, as Italy's first homegrown competitor to the American and British models that commanded the computer market.[33]

A year later, Adriano Olivetti was dead and his family company in need of new leadership. Controversially, a longtime Olivetti executive named Giuseppe Pero was given the nod over Adriano's son, Roberto, who was handed instead the consolation prize of Olivetti's electronics division.[34] The next few years were rocky, to say the least. Underwood was in a more parlous state than anyone had been willing to admit, leading Pero to declare that both factories and working practices would have to be demolished and rebuilt to Olivetti's higher standards.[35]

Olivetti needed something to recapture the old magic. It had always been good at making mechanical devices, such as its typewriters and hand-cranked calculators; now, with the Elea 9003, it was also a player in the highest echelons of the computer age.[36] As Roberto and Tchou discussed what they should do next, the answer seemed obvious: conquer the middle ground. What if a calculator could punch up, or a computer could punch down? Most companies that could afford computers kept a tight rein on them because of their extravagant costs, but the Olivetti pair imagined a device that occupied a desk, not a room, and that could be operated by the average office worker rather than the high priests of the computing department.[37] One might almost call it a *personal computer.*

The effort to build what would become the Programma 101 began in 1962. It was led by one Pier Giorgio Perotto, a protégé of Mario Tchou's, whose team of dapper young technicians* took to calling their machine the "Perottina," after their boss.[38] Tchou, who had died in a car accident in 1961, would not be around to help.[39] But Perotto had been peripherally involved in the development of the Elea 9003 and had absorbed some of the lessons that Mario Tchou had learned along the way.[40] Like the larger machine,

---

* A contemporary photograph shows Perotto and his team in their lab, relaxed and stylish, posing with a prototype of their calculator. They look exactly like actors playing themselves in a biographical movie.

the P101 would be programmable, even if only in a comparatively modest way. It would be designed, not merely engineered, so that its physical appearance and mode of operation were up to the usual Olivetti standards. Perhaps most important, it would eschew the vacuum tubes then used in most computers (and now, courtesy of the Sumlock Anita, in at least one calculator too) in favor of a new type of switch called a transistor.[41]

T HE INVENTION OF the transistor, a device that can reasonably be claimed to have changed the course of human history, was something of an anticlimax. In 1926, a Polish American physicist named Julius E. Lilienfeld had patented a device in which a strong electrical current passing through a piece of copper sulfide could be modulated by a second, weaker current applied to the same block of material.[42] It was the first use of a semiconductor (that is, a material somewhere between a conductive metal and a non-conductive insulator) to amplify an electrical current, but Lilienfeld never completed a practical working version.[43] Even so, the concept was alluring: here was an amplifier with neither the moving parts of a relay nor the fragility of a vacuum tube, and which, comprising as it did only a solid block of a single material, could theoretically be miniaturized almost beyond comprehension.[44]

After decades of research and experimentation across the globe, a trio of Bell Labs physicists succeeded in bringing Lilienfeld's idea to fruition in 1948.[45] For their momentous breakthrough, John Bardeen, Walter Brattain, and William Shockley would be awarded the 1956 Nobel Prize for Physics—yet at the time, the response from the world at large was decidedly muted.[46] A press conference announcing the transistor attracted only a desultory back-page article in the *New York Times*, and industry insiders looked upon the transistor as merely a toy invented by the phone company.[47] Even Bell Labs' parent company, AT&T, was unwilling to bet the farm on this new gadget. Ma Bell used the transistor to replace broken relays and to boost long-distance signals, but that was the limit of her ambition.[48]

Running parallel to this was an ongoing dispute between AT&T and the U.S. government. In 1956, the two parties settled a marathon, seven-year lawsuit that resulted in the break-up of the Bell System of companies; an injunction to stay out of all markets except telecoms; and, crucially, the royalty-free licensing of all AT&T's patents.[49] One of the abiding consequences was that the transistor, released from AT&T's indifferent clutches, made a splash in the field of computing. Where once the vacuum tube had ruled, now the transistor would usurp it—and in a sign of things to come, the United States itself was soon dethroned as the transistor's natural home. By the end of the 1950s, Japan was making more transistors than the United States, and even Italy, once a laggard in such things, had beaten the USA to the punch with the Programma 101—arguably the world's first fully transistorized computer.[50]

THE MACHINE THAT emerged from Perotto's lab was a bruising heavyweight in calculator terms: eighteen inches across, ten high, and twenty-four deep, and weighing in at sixty-five pounds.[51] But heavy or not, the powerful, programmable P101 blurred the lines between calculator and computer. As its users came to appreciate the feats of computation of which the Perottina was capable, the fact that it could fit on a desk at all started to seem like a minor miracle. It was only thanks to the extensive use of transistors, newly liberated from AT&T's patent vault, that the P101 did not require an even bulkier enclosure.[52]

The P101 ingested programs on magnetic cards, fed through a

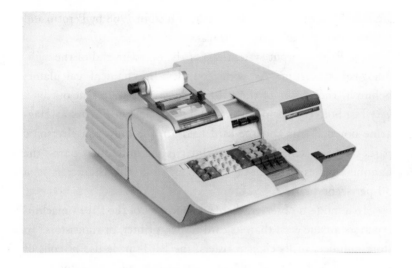

O The Programma 101, engineered by Pier Giorgio Perotto and designed by Mario Bellini. Visible are the keyboard for entering numbers and operations; the central card reader for recording and reading back programs; "V," "W," "Y," and "Z" keys for executing programs, and the printer for displaying inputs and output. © 2012 Alessandro Nassiri | Museo Nazionale della Scienza e della Tecnologia Leonardo da Vinci.

slot above the keyboard, that held up to 120 instructions each. (The cards emerged immediately below a bank of "function" keys labelled "V," "W," "Y," and "Z", and could be annotated by hand to remind the user which program each key would execute.)[53] Programs could add, subtract, multiply, divide, and compute square roots; and the results of those operations could be shuffled between named storage locations called registers. To avoid problems with local translations, the keyboard used symbols rather than words: $\Diamond$ stood for "print"; * meant "clear"; ↑ moved a value to the "M" register; ↓ moved a value to the "A" register; ↕ exchanged the values in the "M" and "A" registers; and so on.[54] Programs could even branch in one direction or another depending on the value of the "A" register.[55] It was heady stuff, and it formed the bulk of a patent filed in 1968 by Perotto and Giovanni De Sandre, one of his team.[56]

The P101's output system was at the opposite end of the technological spectrum. As with countless mechanical calculators before it, the results of the P101's calculations were spat out by a very fast but very analog printer. A 1965 article in *Electronics* magazine noted, "This and the card feed are the only non-electronic parts of the machine, and the printer is the only noisy part—the hammer makes a rat-tat-tat as it bangs out the characters at a 30-per-second rate."[57]

To a modern observer, there was one part of the P101's machinery more arcane even than its card reader, printer, or transistors. To store numbers in its eight registers, the P101 made use of coils of wire called "delay lines," each one plucked like a guitar string by an electromagnet. Pulses took precisely 2.2 milliseconds to travel from one end to the other. When vibrated with a signal corresponding to a number, the P101 waited for the same signal to arrive at the far end of the coil before amplifying and resending it.[58] In this way, the P101's registers hummed ultrasonically with the sounds of the numbers stored within them, circulating endlessly 450 times every second.

THE PROGRAMMA 101 came at an inflection point for design. Mechanical devices, such as traditional calculators, obliged designers to drape an outer skin over an object whose form was dictated by its mechanical function. Electronic components, on the other hand, could be soldered onto circuit boards of any shape or size so that the internal structure of a device could be rearranged to suit its desired exterior form.[59] Designers became architects, in other words, rather than window-dressers. It was appropriate, then, that Olivetti entrusted Mario Bellini, a Milanese architect who had graduated only a few years earlier, with the job of sculpting the boxy P101 prototype into an object worthy of the Olivetti name.[60]

The completed P101 had a foot in both the old world and the new. Its logic circuits were enshrined on a series of circuit boards, but the six-thousand-odd resistors, capacitors, and transistors they held made it impossible to shrink the P101 much beyond the size of a mechanical desk calculator or typewriter.[61] Its bulky electrical components—a motor, magnetic card reader, printer, and power supply—also contributed to its overall volume.[62] Nevertheless, Bellini's design owed as much to his own philosophy as it did to the P101's internal structure: he wanted the calculator to be organic rather than mechanistic and friendly rather than forbidding. To that end, the P101's bright, off-white aluminum casing, itself a far cry from the stamped gray steel typical of mechanical calculators, was punctuated by gill-like cooling vents and a protruding "tongue" that acted as a palm rest. Even the P101's keys were masterpieces of design, each one bearing a raised circular pad that stretched cleanly down to a square base, as if made of rubber or skin.[63] It was a clever detail for something as mundane as a button, and Bellini later used the same motif in another iconic Olivetti calculator, the Divisumma 18, that would become one of his best-known works.[64]

To appreciate the significance of Bellini's design, it is instructive to compare the Programma 101 to two of its closest competitors. First, the gawkily named Mathatronics Mathatron. This programmable printing calculator matched the Olivetti machine's technology and features and pre-dated it by over a year. To its detriment, however, the Mathatron was enclosed in an unlovely case of folded metal that lacked even the blunt charm of the prevailing Brutalist architecture. Its controls were a mishmash of dials and pushbuttons arranged against a grid of desaturated, midcentury colors. Its print head sat exposed on the front fascia of the machine.[65] The Mathatron looked like it belonged in a factory, not an office.

Closer to home, Industria Macchine Elettroniche, an Italian

company, released a fully transistorized calculator in the fall of 1964, yet the IME 84's cutting-edge innards were at odds with its aggressively bland design. A featureless gray casing was enlivened only by a row of Nixie tubes and a few primary-colored buttons.[66] Where the P101 was elegant and sensual, the IME was bluff and standoffish.*

Fifty years later, the Mathatron and the IME 84 appear like artifacts from a forgotten time. Bellini's Programma 101, by contrast, still vibrates with taut lines, thoughtful details, and carefully chosen accents. The Olivetti machine was as revolutionary on the outside as it was on the inside.

IN 1964, as the P101's launch neared, a cash-strapped Olivetti almost lost its crown jewel when the Italian company agreed to sell three-quarters of its electronics division to General Electric.[67] Roberto Olivetti saw the deal as securing the future of his department, but Pier Giorgio Perotto was concerned that the Americans would have no interest in a smaller, cheaper machine like the P101. A visit to GE's lab in Phoenix, Arizona, where Perotto's team were belittled by engineers working exclusively on mainframes, did nothing to convince him otherwise.[68]

The P101 project was saved by two acts of creative subterfuge. First, Perotto did everything he could to present himself as the enfant terrible of the electronics division, aggravating his incoming GE bosses to the extent that he and his team were left behind at Olivetti.[69] Second, a sympathetic executive at the Italian firm—

---

* There must have been something in the air in 1964. Friden's EC-130, Sharp's Compet CS-10A, and Canon's Canola 130, all launched that same year, contended with the IME 84 for the title of earliest transistorized calculator. All suffered from the same aesthetic malaise.

Roberto, perhaps, but the record is not clear—had the P101 sneakily reclassified as a mechanical calculator rather than an electronic one. When General Electric assumed control of the electronics division, only to pull a reverse-Underwood and turn it into a sales organization for GE mainframes, the Programma 101 stayed safely with Olivetti.[70]

This hasty recategorization haunted the P101 for the rest of its life. In modern terms, it is very definitely a calculator, confined to numerical operations and providing only the most rudimentary programming capabilities. Perotto, on the other hand, maintained to his dying day that the P101 was a computer, and at least one Olivetti-Underwood advert agreed, presenting it as "the world's first desktop computer."[71] Certainly, the P101's abilities were so far beyond contemporary calculators as to elevate it to some hazy liminal zone between the two. But perhaps the most perceptive view of the problem was informed by commerce, not semantics. When Hewlett-Packard launched a competing device called the 9100A just a few years later, the company's co-founder Bill Hewlett explained:

> If we had called [the 9100A] a computer, it would have been rejected by our customers' computer gurus because it didn't look like an IBM. We, therefore, decided to call it a calculator and all such nonsense disappeared.[72]

Harking back to Charles Xavier Thomas and his arithmometer, Olivetti debuted the Programma 101 at a grand exposition—in its case, the New York World's Fair of 1964 and 1965. The P101 made its entrance in the summer of the fair's second year, installed in a small room in the bowels of the Olivetti booth so that visitors had to make their way past a battery of mechanical calculators, adding machines, and typewriters before they could lay eyes on Perotto and Bellini's brainchild.[73] Shorn of its electronics division a year earlier, the new order at Olivetti was already making itself felt.

Even sequestered in the depths of the Olivetti stand, the P101

did not stay hidden for long. At $3,200 it was a very expensive calcu-
lator but a very cheap computer. By way of comparison, the Digital
Equipment PDP-8, the first "real" computer sold to end users, had
gone on sale earlier the same year for $18,500. (Adjusted, those prices
are $27,500 and $160,600 respectively. "Cheap" is relative.)[74] Indeed,
enthralled onlookers could not quite believe that the Programma 101
was a standalone device. They looked for a cable connecting it to the
computer they assumed must be hidden elsewhere, but they did not
find one.[75]

The Programma 101 was also the first Olivetti product launched
not in its home country but rather in the computer-hungry United
States.[76] The decision was amply rewarded: the P101 would go on to
sell 44,000 units during its lifetime, with 9 out of every 10 going to
American customers.[77] Olivetti had imagined the P101 being used
in business, engineering, banking, and science—and to be sure, it
found customers in all of those industries—but its inherent power
and flexibility also led to roles in two of the most consequential
American projects of the decade.[78]

First, the Programma 101 joined the race to the moon. NASA's
engineers found the P101 to be perfect for one very specific job, as
related in a 2006 interview with David W. Whittle, the flight con-
troller in charge of the communications equipment installed aboard
the Apollo 11 lunar module:

> [T]he Lunar Module high-gain antenna was not very smart.
> It didn't know where Earth was. So you would have to call up
> and give the astronauts some—we had two knobs, a pitch and
> yaw knob, but you have to give him some angles to put it at.
> [. . .] We would have to run four separate programs on this Pro-
> gramma 101, and then in between those programs, we'd have to
> get out our manuals [. . .] we'd have to look up trigonometric
> functions and input the data, which today your calculator does

that. [. . .] Then we would read out the angles that we came up with to the crew, and they would dial them in, look at the signal strength, the signal strength there.[79]

Olivetti's "supercalculator," as Whittle called it, was a vital cog in the effort to keep Neil Armstrong and Buzz Aldrin in touch with their home planet.

Shortly thereafter, the Programma 101 was pressed into the service of the *other* fateful American adventure of the 1960s. The same year that Aldrin and Armstrong landed on the moon, the U.S. Air Force began secret bombing operations over neutral Cambodia, seeking to dislodge North Vietnamese forces from hidden Cambodian bases. Coordinates for these raids were sent to the U.S. airbase at Biên Hòa in Vietnam, where radar operators used Programma 101s to compute ranges, bearings, and bombing trajectories for the B-52s that would be sent across the Cambodian border after nightfall.[80]

A BLOCKBUSTING DEBUT in New York City; a ready-made U.S. sales organization; customers at the most elevated and the grubbiest extremes of the American establishment: whether you called it a calculator or a computer, the Programma 101 had America at its feet. Where did it all go wrong?

The most common verdict was death by a thousand cuts. The Underwood deal had been an expensive gamble, and the construction in 1968 of a stylish new Olivetti-Underwood plant in Pennsylvania ate into the payoff.[81] The same year, the P101 encountered its first real competition in the form of Hewlett-Packard's 9100A, a clone so brazen that if HP had stood on the shoulders of giants they must have been picking their pockets at the same time.[82] The American company would ultimately pay Olivetti almost a million dollars for infringing Perotto and De Sandre's patent, but the damage was done and HP's rival machine was there to stay.[83]

Not helping matters was the Italian government's apathy towards the new science of computing. Prior to 1954, barely a dozen Italians had gained any experience in the field, and even after Olivetti hired its way to competence and gifted an Elea 9003 to the Italian Treasury, there was no state aid to be had.[84] It was a stark and galling contrast to the support offered to British and American computer companies—and, speaking of which, Olivetti was still reeling from the hasty divestment of its electronics division to America's General Electric.

Then there were the rumors.

A certain shiftiness swirled around Olivetti during this most turbulent period. Adriano had made no secret of his desire to sell the Elea 9003 in communist China, and to some, his untimely death from a heart attack could not be coincidental. Carlo De Benedetti, later Olivetti's CEO, was of the opinion that IBM had done away with his predecessor with the aid of U.S. intelligence services, who, after all, had had Adriano on their radar since the Second World War.[85] And if one subscribed to that theory, it would have been doubly unnerving to learn of the death only a year later of Mario Tchou, the wunderkind designer of the 9003, in a freak car accident. In 1961, Tchou and Roberto Olivetti had got as far as Hong Kong before visa issues forced them to turn back from a planned visit to Beijing, and the engineer was dead just days after arriving back in Ivrea.[86] As if to provide a finale to a trilogy of suspicious events, the sole prototype of the Programma 101 was stolen from Ivrea shortly before its New York debut and spirited across the Alps toward unknown buyers in Switzerland. Ivrea's chief of police got wind of the heist from an informant and intercepted the thieves, but little else is known about the incident—even if it is not hard to read a lot more into it.[87]

Buffeted by commercial missteps, governmental indifference, and alleged espionage of both the industrial and state-sponsored varieties, Olivetti failed to capitalize on the success of the Programma 101. Who knows? Had Olivetti been managed better, had it not lost its charismatic leader and best engineer in short order, perhaps the Casio or Texas Instruments calculator in your desk drawer could have been an Olivetti instead.

10 THE TEXAS INSTRUMENTS CAL TECH

PAT HAGGERTY, the president of Texas Instruments, was not a man to waste an opportunity. In the 1950s he had pioneered the pocket transistor radio as a way to sell transistors, and now, a decade later, Haggerty needed a vehicle for TI's latest product. The company already supplied the integrated circuits, or "microchips," used in the USA's Minuteman nuclear missiles, but such was the destructive power of each individual warhead that the market for doomsday missiles was necessarily a small one. On a flight to Dallas in the summer of 1965, Haggerty badgered Jack Kilby, the inventor of the integrated chip and one of TI's star engineers, to design something, *anything*, with which TI could sell microchips to the average American consumer.

JACK KILBY JOINED Texas Instruments in the summer of 1958. He had been hired to work on products called micromodules—electronic components of identical size and shape that could be assembled, LEGO-fashion, into bespoke circuits—but Kilby was skeptical of the idea, having seen similar projects stall out in both military and private spheres. Now, with no accumulated vacation time, he had TI's semiconductor lab to himself while his new colleagues sunned themselves elsewhere.[1] And fearing a future wasted on pointless micromodule projects, he set himself a goal: come up with a better alternative before the holidays were over.[2]

At the dawn of electronics, when telegraph circuits and radio transmitters comprised only a few components apiece, it had been easy enough to wire them up by hand. By the mid-twentieth century, however, electronic devices had become much more intricate and extensive, and none more so than those used by America's armed forces. As a case in point, each of the U.S. Air Force's four thousand B-29 bombers carried a fire-control computer that contained a thousand vacuum tubes.[3] Manufacturing such complex circuits was a time-consuming and error-prone process, and throughout the twentieth century, a series of engineers and inventors had tried to make electronic circuitry more amenable to mass production.[4]

It began with a Prussian inventor named Albert P. Hanson, who filed a patent in 1903 that set the stage for everything that would follow. Hanson described how to stamp a circuit out of metal foil and sandwich it between layers of protective, insulating paper. Holes in the paper would permit electronic components to be affixed to the conductive metal below. In a prophetic closing remark, Hanson wrote that one day it might be possible to make circuits merely "by printing with metallic powder in a suitable medium."[5]

It took decades for others to turn Hanson's ideas into a practical reality, a journey that included along the way a not insignificant flip from the Prussian's "additive" method, where a circuit is laid onto an insulating substrate, to "subtractive" methods, where a conductive plate is etched away with acid to form the desired layout. Even so, Hanson can be credited for inventing what has become known as the printed circuit board, or PCB, an innovation that has underpinned electronic devices, literally and figuratively, since the Second World War.[6]

PCBs helped designers build ever more sophisticated circuits and encouraged them to imagine even greater things. Desktop devices like Sumlock's Anita and Olivetti's Programma 101, for instance, relied on PCBs to tame their thousands of components. But when engineers imagined the sort of computers that might guide a rocket to the moon—computers that would need not thousands but millions of components—PCBs were of little help. Designing the circuitry for a moon rocket's computer was one thing; fitting it inside that rocket was quite another. Engineers called it the tyranny of numbers, and for years they had been trying to break it.[7] Jack Kilby, during that first summer in the deserted lab at Texas Instruments, would be the first to succeed.

TEXAS INSTRUMENTS had been founded in 1930 to prospect for oil. Then called Geophysical Service, Inc., its founders had developed a way to use sound waves to map the kinds of subterranean features that harbored deposits of crude oil, and, in doing so, had put themselves at the forefront of electronic design. It transpired that their method for searching for oil in the ground could be applied equally well to hunting planes in the sky or submarines in the sea, and the Second World War turned GSI's business on its head. Electronics overtook oil, and in 1951, the firm was reborn as Texas Instruments.[8]

All the while, the state of the art in electronics was advancing in leaps and bounds. Pat Haggerty, a Navy lieutenant who had bought GSI equipment during the war and who joined the company after it, was anxious to keep up, and in 1953, he poached from Bell Labs a semiconductor expert named Gordon Teal.[9] In a twenty-year career at Ma Bell, Teal had learned how to grow extremely pure silicon crystals and also how to "dope" them with impurities to control their conductive properties. His research promised to revolutionize electronics: germanium, the element from which John Bardeen, Walter Brattain, and William Shockley had made their transistors, and which still dominated semiconductor manufacture, was difficult to purify and yielded components that worked only within a narrow band of temperatures. In theory at least, Teal's putative silicon transistors would suffer neither such problem.[10] He pushed forward with this work at Texas Instruments.[11]

All this led to a now-legendary presentation at a 1954 conference on electronics. There, Teal declared that Texas Instruments could manufacture three distinct varieties of silicon transistor, a feat that had singularly eluded the rest of the industry, and held up a handful of such components to prove it. Next, he brought the house down with a demonstration. Firing up a record player amplified by germanium transistors, he immersed one of them in a container of hot oil and, to no one's surprise, the music cut out instantly. He repeated the feat with a silicon transistor, and the music played on, even as the transistor bathed in the shimmering heat of the oil. There was an unseemly rush to grab copies of his conference paper from the back of the room: Teal's audience had just witnessed the defenestration of expensive, unreliable germanium by cheap, robust silicon. A new semiconductor age was at hand.[12]

Prior to this demo, Teal had hired a chemist named Willis Adcock to help TI move from finicky germanium to more robust silicon; Adcock, in turn, had convinced Jack Kilby to join the company to explore what else might be done with doped silicon.[13] Kilby's plan, as recorded in his notebook on July 24, 1958, was deceptively straightforward.[14] It was possible to make a host of electronic components from little chunks of doped silicon: resistors to control current, diodes to dam it, capacitors to store charge, and transistors to switch or amplify.

But why bother making individual components, Kilby reasoned, when you could bake them all into a single slice of silicon?[15] Adcock was skeptical, but not so skeptical as to forbid Kilby to try it out.

Ironically, for all TI's vaunted expertise in silicon, Kilby could not find a large enough piece of it. Instead, he and an unnamed female colleague ("a girl," he called her in a 1975 interview) fell back to a slab of germanium that they carefully etched and doped so that it contained all the components of a simple circuit called a phase-shift oscillator. He finished the circuit with the judicious application of connective wires, and, on September 12, 1958, he hooked it up to a battery and a monitoring device called an oscilloscope. The oscilloscope's screen lit up with a sine wave—the telltale signal emitted by the world's first integrated chip.[*],[16]

Kilby's own moment in the sun came in the spring of 1959 at the Institute of Radio Engineers' annual convention. There, Pat Haggerty announced that Texas Instruments could now emplace entire circuits on chips of silicon just a fraction of an inch in size. It should have been a fitting sequel to Gordon Teal's revelation of the silicon transistor, but Kilby's invention attracted debate rather than adulation. Silicon was the perfect material for individual transistors, but detractors noted that crowding them onto a block of the stuff with a host of neighboring components noticeably reduced their performance. To compound matters, it took considerable time and effort to bring those carefully laid-out microchips from design to production, piling pressure on designers to get things right the first time. And even once a chip's kinks had been worked out, fabricating it was a still a crapshoot: thanks to the vagaries of early production methods, only twelve out of every hundred copies of a chip bearing

---

* Kilby's second test circuit was a flip-flop, the classic switch-based memory circuit first invented in 1919, but with vacuum tubes replaced by transistors.

just twenty embedded components might be expected to survive the manufacturing process.

"These objections were difficult to overcome," Kilby deadpanned in a paper describing his invention, "because they were all true."[17]

YET FOR TEXAS INSTRUMENTS, the promise of the integrated circuit* outweighed the disadvantages, and the firm pushed ahead regardless.[18] In 1958, the same year that Jack Kilby had pioneered the integrated chip, the U.S. Air Force had set itself the goal of making electronics circuits ten times smaller than current systems while increasing reliability by the same factor.[19] Of all the companies that made up the latter component of the military-industrial complex, Texas Instruments was the first to meet the Air Force's challenge, and it did so in typically showy style. In October 1961, Pat Haggerty joined Air Force personnel to demonstrate two computers to audiences in Dayton, Ohio; Washington, DC; and Los Angeles, California.[20] The two machines were functionally identical, capable of little more than addition, subtraction, multiplication, division, and square roots, but in construction they were radically different. The first was built from more than nine thousand discrete components and weighed about the same as two cases of beer; the second contained a mere 587 integrated chips and weighed less than a single bottle of the stuff.[21]

The firm grabbed another headline the same year when *Fortune* magazine featured TI's new line of integrated circuits on the cover. Each of TI's "Series 51" chips contained a simple electronic network,

---

* In these early years, the jargon was still fluid. A lump of silicon or germanium that hosted a collection of components might be called a chip, a microchip, a slice, a wafer, or an integrated circuit.

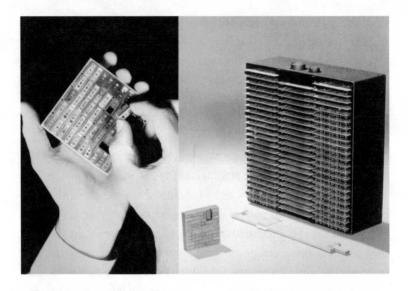

○ Texas Instruments' dueling computers, completed in 1961 for the U.S. Air Force.
On the left, the integrated-chip version, small enough to fit in the hand. On the right,
its hulking counterpart, made using discrete components, with a slide rule for scale.
SMU Digital Collections. "First IC Computer." Southern Methodist University, 1961.
https://digitalcollections.smu.edu/digital/collection/tir/id/14. Image creator Texas
Instruments, part of the Texas Instruments records, DeGolyer Library, SMU.

such as a flip-flop memory circuit, that could be combined with
others to carry out more sophisticated operations.[22] Two years later,
Series 51 chips became the first integrated circuits to go into space
aboard NASA's *Explorer 18* satellite.[23] And while the Air Force's
nuclear-tipped Minuteman II missiles made for a more sinister application than NASA's innocuous space physics mission, TI did not
hesitate to mention the use of their microchips within the missile's
guidance systems at every available opportunity.[24]

Privately, TI execs were worried about the lack of demand
for microchips beyond their institutional customers. In 1965, for
example, NASA published data showing that the integrated cir-

cuits used in the Apollo space program could be expected to last for up to 20 million hours, and who would ever need to buy a second radio or television if their first one kept working for two thousand years or more? [25] Equally, computer manufacturers remained unconvinced of the need to upgrade from transistors to microchips, despite the obvious benefits of smaller, cooler components.[26]

Pat Haggerty had seen this problem before. He had solved it too.

Back in 1951, TI had developed Bell Labs' germanium transistor designs into a saleable product line, but faced an uphill struggle to sell transistors in bulk. To the extent that the average person on the street knew or cared about transistors, they did not trust them. Worse, neither did the manufacturers that made computers, radios, televisions, and similar devices. [27] Vacuum tubes were just fine, thank you. In 1954, then, Haggerty formulated a plan to take the transistor out of the lab and into the public consciousness. In May of that year, he gave TI's engineers a budget of $2 million to develop a miniature radio built around the company's transistors, and brokered a deal with a consumer electronics manufacturer to make and sell it.[28] At Christmas in 1954, less than a year after the project had started, the pocket-sized Regency TR-1 transistor radio was launched to great acclaim and would go on to sell a hundred thousand units in its first year. Consumers were convinced, and IBM and other corporate refuseniks soon followed.[29]

In 1965, Pat Haggerty decided to try the same trick for the integrated circuit—to conjure up a product that the average American consumer might buy, and thereby plant the idea that microchips might not be so bad after all. On that fateful flight to Dallas with Jack Kilby, the two conferred over the best way forward. A lipstick-sized Dictaphone, Haggerty wondered? Or, perhaps, a calculator that could fit in one's pocket?[30]

o——o

O The Regency TR-1 AM radio, driven by Texas Instruments' transistors. The TR-1 sold for $50, or about a week's wage for a working adult.[31] "Regency Model TR-1 Transistor Radio." National Museum of American History. Washington, DC: Smithsonian Institution, 1954. https://americanhistory.si.edu/collections/search/object/nmah_713528. Image courtesy of Division of Medicine and Science, National Museum of American History, Smithsonian Institution.

HAGGERTY AND KILBY'S PROJECT arrived at a tumultuous point in the history of the calculator. In the two decades since the end of the war, the received wisdom on how to build calculators had been overturned not once but four times. Stolid desktop machines, such as Katherine Johnson's STW-10, had been surpassed first by Curt Herzstark's pocketable Curta and then by Casio's relay-based 14-A. In short order, the Sumlock Anita replaced the Casio's clattering relays with the fragile magic of the vacuum tube. Finally, transistorized calculators such as Friden's EC-130, Sharp's Compet CS-10A, and the IME 84 tumbled over one another in an unseemly rush to capitalize on the promise of the transistor.[32] Only a decade before, calculators had been big, slow, and noisy; now, anyone with enough money to spend could buy a calculator that did away with any two of those foibles. Courtesy of the microchip, Haggerty and Kilby were trying to kill all three birds with a single stone.

In September 1965, Jack Kilby invited a group of engineers to his office and pointed to a small book on his desk. We are going to figure out how to make a calculator the same size, he told them, with silicon chips for math, buttons for input, batteries for power, and a display of some kind for the answers. Or rather, he told them, *you* are going to figure out how to make this calculator.[33] Among the group was a high-flyer named Jerry Merryman, a tinkerer with electronic gadgets since his elementary school days who had a flair for intricate, clever circuits.[34] Less than two years into his tenure at TI, Merryman had already made his name with a high-end memory circuit called a shift register, which placed 352 interconnected components on a single silicon chip. An image of Merryman's design graced the cover of TI's annual report for the year 1965.[35] For the calculator, Merryman sketched out in three days a circuit that could add, subtract, multiply and divide—and then, after being picked by Kilby to lead the project, spent two long years figuring out how to get that same design onto a set of integrated chips.[36]

Daringly, Merryman took his designs straight from the drawing board and embedded them in four large integrated circuits of around 150 components each. The conventional approach to developing a new chip was to wire up a prototype from discrete components mounted on, say, a wooden backing board, and *then* figure out how to arrange those components on a silicon chip. Marching straight into a chip foundry with a blueprint under one's arm and hope in one's heart was unheard of. Worse, Merryman's chips were an order of magnitude more complex than TI's usual fare. If the success rate when manufacturing a chip with 20 components was typically 12 percent or less, how on earth could TI ever hope to mass-produce chips bearing 150 components each?[37]

And yet, Merryman's approach worked. His chips were complex but also conservatively designed: their components were widely spaced to improve manufacturing yields, and each chip included a handful of redundant components in case others did not make it through the fabrication process. When the first few chips did not work, Merryman reluctantly constructed a "breadboard" model to figure out where the defective components lay, and peering carefully through a microscope, he manually rewired the chips around them.[38]

As work on the chips progressed, TI's accountants started to ask for a name against which project expenses could be recorded. By tradition, projects were code named for universities and other educational institutions, and so the "Cal Tech" project was born, a nod to the California Institute of Technology.[39] Whatever good vibes might have been gained from associating with such an august institution, they evaporated when Kilby realized that "Cal Tech" was uncannily descriptive of the project's goal to advance the state of *cal*culator *tech*nology.[40]

Kilby himself addressed the calculator's power pack, which comprised novel, rechargeable nickel-cadmium batteries, and recruited an engineer named James Van Tassel to tackle the calculator's keypad.[41] Very little has been said about this aspect of the Cal Tech's design, but Kilby, Van Tassel, and Merryman must have realized that a hundred-key comptometer keyboard would never fit. Instead, they chose the ten-key numeric keypad layout that had first appeared on adding machines toward the end of the nineteenth century, and which had since begun to overhaul comptometer-style layouts even on full-scale desktop calculators.[42] Certainly, by 1960, when engineers at Bell Labs were developing the first pushbutton telephones, they could cite the ten-key calculator keypad as an inspiration for their chosen design.[43]

The final part of the Cal Tech puzzle was the display mechanism, the only part for which there was not an obvious solution. Nixie tubes, those tiny bulbs with electrodes bent into the shape of numeric digits, were too power hungry. And though, in the early 1960s, engineers at Texas Instruments had invented a new component called a light-emitting diode, LEDs were still too expensive for the calculator project.[44] Instead, the answer came from yet other engineers at TI, who had married the space-age technology of the silicon chip to the ancient craft of printing. Their system hinged on a microchip bearing fifteen tiny resistors, each of which grew hot on receiving a current, and that could be activated in different patterns to "burn" tiny dot-matrix characters onto a reel of heat-sensitive paper.[45] The printer used power only while printing, and the hard copy it pro-

O   The Cal Tech prototype, completed late in 1966 and presented to Texas Instruments President Pat Haggerty the following year. "*Handheld Electronic Calculator Prototype: Texas Instruments Cal Tech.*" *National Museum of American History. Washington, DC: Smithsonian Institution, 1967. https://americanhistory. si.edu/collections/search/object/nmah_1329686. Image courtesy of Division of Medicine and Science, National Museum of American History, Smithsonian Institution.*

duced would be invaluable for accountants and others who needed a permanent record of their work. It got the nod.[46]

By November 1966, scarcely a year after the team had started their work, the Cal Tech prototype was ready. The electronics, batteries, keypad, printer, and reel of paper were packed into a hollowed-out piece of aluminum six inches tall, four across, and an inch and a half deep. It was about the same size as a large paperback novel, although at almost three pounds in weight, or more than a kilogram, it weighed a lot more.[47] To call the Cal Tech "pocket-sized" was to declare that you owned at least one unusually large and robust pocket, but it was, nevertheless, an order of magnitude smaller than the desktop behemoths that dominated shop windows and offices. Haggerty was satisfied.[48]

There was, however, a problem. TI's chip fabrication techniques had come on in leaps and bounds since Jack Kilby's early experiments. Rather than painstakingly etch individual resistors and transistors into raw blocks of silicon, now the company made tens of copies of each circuit at the same time on a single inch-square wafer of silicon and then cut them apart to keep the ones that worked. But though Jerry Merryman had designed the Cal Tech's circuits with considerable margin for error, the company could not produce his giant, half-inch chips in bulk. For two years, the project stalled.[49]

EVERY NOW AND AGAIN, an industry lurches toward a new center of gravity. For American drivers, the arrival in the 1950s of imports such as the Volkswagen Beetle signaled that Detroit no longer had the United States to itself.[50] A decade later, the Beatles (no relation) were the vanguard of a British invasion that had both teenagers and record labels looking to the other side of the Atlantic.[51] And as Texas Instruments' movers and shakers vacillated over what to do about Jerry Merryman's microchips, they watched a similarly seismic shift engulf the electronics industry. Imports from Japan were on the rise, and consumers were going to have to make their peace with gizmos made by their wartime foes.

Like Italy, postwar Japan had been the beneficiary of an unprecedented "economic miracle." Unlike that of Italy, however, Japan's recovery had been carefully planned and vigorously pursued by the Japanese state. The government rushed to the aid of industries it saw as the foundations of a successful, modern economy, such as car making, steel production, and semiconductor electronics.[52] Foreign companies were allowed into the country only through joint ventures with local firms, while the government encouraged those local enterprises to ally with one another and funded them with favorable loans.[53]

Hand in hand with these protectionist measures went a new focus on quality and design in manufacturing. W. Edwards Deming, a statistician who had preached a gospel of quality control to unreceptive American ears, found a more eager audience in Japan.[54] And from 1959 onward, the Japanese government began to take seriously accusations of plagiarism in product design.[55] Gradually, the stigma of the "made in Japan" label was erased.

And yet, the future of Japan's electronics industry was decidedly equivocal. By the early 1960s, the country had overtaken the United States as the world's largest producer of transistors, but it lagged in its ability to design and manufacture integrated chips.[56] This led to a series of alliances between those American firms who could make the chips and the Japanese companies who could assemble them into finished products in an economically viable way. Sharp partnered with Rockwell; Sanyo with General Instruments; Casio with Fairchild; Busicom, later, with Intel. But TI's presence in Japan was felt more often in legal disputes than in products.[57]

○——○

A S EARLY AS 1960, TI had applied to Japan's Ministry of International Trade and Industry (MITI) for patent protection on the integrated circuit.[58] Those patent applications were still in limbo four years later, when TI asked MITI for permission to open a Japanese subsidiary. Negotiations dragged on for years, reaching a nadir in 1966 with public complaints about Japan's stubborn protectionism from the U.S. commerce secretary, the newly formed international Organisation for Economic Co-operation and Development (OECD), and a host of foreign governments.[59]

And as TI fulminated, Japanese companies innovated. June 1967 saw the launch of Sony's "Sobax" calculator, its name a portmanteau of "solid state abacus," and which was driven by discrete components cleverly packaged to reduce their footprint. The Sobax was the first electronic calculator to have meaningful concessions to portability, in the form of a handle on top and the ability to draw power from an external battery pack or a car's cigarette lighter.[60] At more than a thousand dollars in price and fourteen pounds in weight, no one would mistake it for a pocket calculator, but it was a sign that smaller, lighter devices were on the way.[61]

A much more serious competitor for the Cal Tech emerged late in 1969. Sharp's collaboration with Rockwell had resulted in the QT-8D, or "Micro Compet," the first calculator to use a new type of large-scale integrated chip that each contained upward of nine hundred individual components. At a svelte three pounds in weight and five by ten inches in size, the plug-in Micro Compet was perilously close to Cal Tech territory. Its battery-powered sibling, the QT-8B of 1970, even more so.[62]

Finally, in 1968, TI settled its dispute with MITI. In exchange for permission to open a semiconductor factory in partnership with Sony, TI agreed to license its valuable integrated chip patents to Japanese manufacturers.[63] Just in time too: by the turn of the decade, calculator manufacturers were Japan's single biggest consumer of integrated chips.[64]

Nor had TI been idle at home. By 1968, improvements in the company's chip-making process had allowed TI to make Jerry Merryman's Cal Tech chips smaller and cheaper.[65] The following year, after quietly shopping their prototype calculator to potential partners, TI signed a deal with Canon to produce and sell the Cal Tech to consumers. Five years after Jack Kilby and Pat Haggerty had first decided to build a calculator to sell more microchips, they could finally go ahead and do exactly that.[66]

AS WITH MANY electronics companies in Japan and elsewhere, Canon had got into the business by the side door. TI had been founded by oil prospectors; the Kashio brothers had made finger rings for nicotine addicts; Sumlock had made ticket punches. Canon, for its part, was a camera company by choice and a calculator company by coincidence.

In the early 1960s, Canon's high-end still cameras came with integrated light meters and electronic flash synchronization, and their TV cameras were essentially nothing *but* electronics.[67] The company had lucked into a fortuitous position: to evolve their cameras, Canon's engineers needed to solve ever-more-complicated optical equations, but they had also acquired enough in-house electronics expertise to design a calculator with which to tackle such thorny math problems. The result, a transistorized desktop calculator called the Canola 130, was announced in prototype form in 1964 and gave rise to a family of similar machines.[68] By 1970, Canon had settled into a respectable second place behind Sharp, producing around 1 in 7 calculators made in Japan.[69]

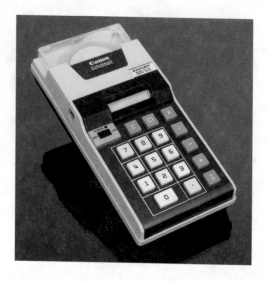

O The Pocketronic, Canon's user-friendly, mass-production version of the blocky Cal Tech. *CC BY-SA 2.0 image courtesy of Vicente Zorrilla Palau.*

Now handed the job of bringing the Cal Tech to market, Canon refined TI's blocky aluminum prototype into a machine with a friendly plastic casing and a magnifying lens over its print head. Using TI microchips, the calculator could churn through thousands of operations during its three-hour run time, during which its user might expect to use up most of the 250 feet of heat-sensitive tape contained within a removable plastic cartridge.[70]

Announced on April 14, 1970, the day before U.S. taxpayers were due to file their returns, the Canon Pocketronic was the first calculator that anyone had dared to place in conjunction with the word "pocket." It was deserved, just about, since Canon's engineers had shaved a pound off the prototype's forty-five-ounce heft, although at eight inches by four it kept the Cal Tech's none-too-dainty footprint.[71] In other ways, too, the Pocketronic failed to excite. Its solid-state printer was innovative, certainly, but looked anachronistic when compared to the glowing Nixie tubes and fluorescent displays that featured on many of its competitors. Internally, too, the Pocketronic was already behind the times, since integrated circuits were no longer the novelty they had once been.[72] Accordingly, the debut of the Pocketronic drew polite acknowledgment rather than effusive praise. *Popular Science* reported on Canon's new $400 calculator in August 1970 but, on the same page, lavished twice as many column inches on a yacht that cost ten times more.[73]

And then the manufacturing problems resurfaced. TI had difficulty bringing the Pocketronic's print head to full production until 1971, by which time the company's lead had been almost entirely squandered.[74] When *Popular Science* revisited the Pocketronic, it placed it alongside an equally compact model from Sanyo and Sharp's even smaller EL-8.[75] The *New York Times* would not mention the Pocketronic until 1972, and it could find little to say except that Canon's "somewhat bulky pocket model" was distinguished by having a printer rather than a display. Two years after its introduction, the Pocketronic had already acquired a whiff of obsolescence.[76]

WAS THE FIRST "pocket" calculator a failure? Certainly, the Pocketronic was so late in seeing the light of day that the impact of Kilby, Merryman, and Van Tassel's remarkable work was diluted by the many competitors who had since pulled off the same trick. From a narrower perspective, however, the Cal Tech project was a resounding success. The Pocketronic might not have set the world alight, but TI had gained valuable experience in designing and making complex chips *and* had successfully crowbarred its way past Japan's forbidding trade barriers. TI's tie-up with Canon would lead to orders for many more integrated chips, and sooner than anyone might have realized, the American company would gain the confidence to strike out on its own—a decision that would have stark consequences for Canon and many others. Texas Instruments and the pocket calculator had unfinished business.

11 THE BUSICOM 141-PF

Y THE EARLY 1970S, the calculator was as hot as bell bottoms were wide. The Pocketronic, the long-delayed fruit of Texas Instruments' and Canon's trans-pacific tie-up, was the first device to show the way toward a pocket-sized future, and a host of lookalikes soon followed. TI, Rockwell, and others squeezed more and more transistors onto each integrated chip, while Casio, Sharp, and Sanyo put those chips into a tidal wave of new calculators. On the fringes of this tsunami of innovation lay two niche players—one old, one new, both tiny—who would gate-crash the calculator party in some style. In doing so, and almost by accident, they would sow the seeds of the calculator's downfall and open the door to a new world of ubiquitous computing.

○———○

ROBERT NOYCE WAS a legend of the semiconductor indus-try, a near-mythical figure who had learned at the feet of William Shockley, one of the inventors of the transistor, and who, in 1957, had been one of the "traitorous eight" engineers who left

Shockley's dysfunctional, eponymous company and struck out on their own. In the ten years since, Noyce and his co-conspirators had turned their new enterprise, Fairchild Semiconductor, into one of the most innovative and influential players in the business—and, latterly, they had watched in dismay as it, in turn, was stricken by defections and losses. In 1968, Noyce abandoned ship for the second time, taking with him only Fairchild's research director, Gordon Moore.[1] At their new company, based in California's Santa Clara Valley, the pair planned to make transistorized memory chips. Their company's name was a contraction of "integrated electronics": Intel was born.[2]

It was a difficult time for American chip makers. Military funding was drying up and IBM, the biggest buyer of microchips in the United States, had started to manufacture its own.[3] Noyce and others turned instead to Japan, whose consumer-electronics firms had developed a voracious hunger for integrated chips. And so, in November 1968, Noyce visited one Tadashi Sasaki of Sharp Electronics in the hope of winning some business for his three-month-old company.[4]

Sasaki was happy to receive the Intel delegation.[5] Jack Kilby might have worked out how to place multiple components on a single silicon chip, but the surfaces of his microchips were crisscrossed by forests of manually soldered wires. Then, in 1959, Fairchild's Robert Noyce combined two breakthroughs—an improvement in chip fabrication pioneered by his colleague Jean Hoerni, and his own novel process* for depositing electrical circuits directly onto silicon chips— to transform the integrated chip from a hand-soldered curiosity to a mass-produced commodity.[6] The Japanese semiconductor industry leaned heavily on Noyce's "planar" technique, and Sasaki's own rise at Sharp owed much to his licensing of it.[7]

"Rocket" Sasaki was a formidable figure in his own right.[8] He had

---

* Proving that there is nothing new under the sun, Noyce's method of laying an electrical circuit directly onto a silicon chip had much in common with the concept of the printed circuit first articulated by Albert Hanson in 1903.

persuaded Rockwell, an American semiconductor firm, to gamble on a new type of low-power, high-density microchip and had negotiated an exclusive contract to buy Rockwell's production of such chips.[9] These metal-oxide semiconductor large-scale integrated chips— MOS LSI, for short—powered Sharp's handheld QT-8D and QT-8B calculators and helped make them persuasive alternatives to Canon's Pocketronic.[10] Sasaki would go on to become a doyen of Japan's electronics industry, with both Apple's Steve Jobs and Masayoshi Son of Japan's giant SoftBank conglomerate happy to call him a mentor.[11]

For now, though, Sasaki had bad news for his guests. Sharp's deal with Rockwell forbade the Japanese company from buying chips from anyone else, even the one and only Robert Noyce.[12] But perhaps Sasaki could help in another way. He had been keeping a close eye on the fortunes of Busicom, a much smaller calculator maker led by a fellow alumnus of Kyoto University, and he thought that Busicom and Intel might do business together. He made the necessary introductions.[13]

THE NIPPON CALCULATING Machine Corporation had started out making hand-operated mechanical calculators in 1945. They stuck with them, too, for an extended period: as late as 1967, NCM would launch a hand-cranked model, called the HL-21, whose operator's manual opined that "computations can be made in a comfortable mood thanks to the rise of a pleasant rhythm from the quiet operating sounds and the smooth mechanical revolutions."[14]

The HL-21 was a dinosaur, but NCM knew this. It was not for nothing that NCM had recently rebranded itself as the "Business Computer Corporation," or Busicom.[15] And if Busicom had been reluctant at first to move away from mechanical calculators, it now jumped into the new world of electronics with both feet. Busicom built its first solid-state calculator from discrete components in 1966, before adopting Fairchild chips the following year *and* licensing a series of designs for programmable calculators from a small American firm named Wyle Laboratories.[16] A parade of electronic calculators followed, keeping pace, more or less, with Japan's calculator giants: Sharp, Casio, and Sanyo.[17]

That changed late in 1970, when Busicom quietly released a new version of its "Junior" calculator. The Junior was a small, bland desktop calculator, encased in beige and black plastic and capable only of the four basic arithmetic operations.[18] If the Junior was notable for anything at all, it was the row of glowing blue-green digits that formed its display. These were vacuum fluorescent displays, or VFDs, whose segmented electrodes emitted light at much lower voltages than hot, hungry Nixie tubes.[19] Invented by Sharp mostly to get around Nixie tube licensing fees, VFDs' ghostly digits first appeared on a Sharp calculator in 1967; by the time the Busicom Junior arrived in 1970, they were on their way to becoming a familiar part of calculator anatomy.[20] (VFDs would eventually invade every corner of consumer electronics, from calculators to home stereos to car dashboards.) But the Junior's unremarkable appearance was deceptive.

As Busicom fought to keep pace with the rapidly evolving calculator market, its execs formed a two-pronged plan of attack. First, knowing that America's semiconductor manufacturers were working to reduce the number of chips required to build a calculator (fewer chips meant quicker, cheaper assembly), Busicom decided to leapfrog ahead to the logical conclusion of that race. It wanted a single chip that could power all its low-end, consumer-oriented calculators.[21]

Busicom found a partner in the form of a U.S. chip maker called Mostek, then just two years old and a minnow compared to giants such as Fairchild, Rockwell, and Texas Instruments.[22] Yet Mostek did what those giants could not. In November 1970, six months after signing a contract with the Japanese firm, Mostek delivered a chip designed to Busicom's specifications that measured one-fifth of a square inch, contained more than two thousand transistors, and could add, subtract, multiply, and divide numbers up to twelve digits long. Mostek's competitors were left in the dust: American Micro-systems, its closest rival, had managed only a two-chip calculator system, while the three oversized wafers of Texas Instruments' calculator chipset were a testament to that company's ongoing production problems.[23]

The updated Junior, which surfaced late in 1970 or early in 1971 (the exact date is not clear), was almost identical to its predecessor. A discerning user might have noticed that the new model was a fraction leaner than the old one, and might also have detected a faint sluggishness in the new Junior's cogitation. The switch from the twenty-two separate chips of the original to the "calculator on a chip" of the new version might have slimmed down the Junior, but issues with early MOS LSI chips dulled its reactions into the bargain.[24]

The Junior's subtle refresh was a warm-up for a much bolder move. The Busicom Handy-LE, the world's first pocket calculator, broke cover in March 1971.[25]

It was inevitable, really. Sharp had fired the first shot with the compact QT-8D, and Canon had replied with the Pocketronic. Chips were getting smarter and calculators were getting smaller, and it was only a matter of time before a pocket calculator worthy of the name would emerge. That calculator was Busicom's LE-120A, or Handy-LE. Two and a half inches across, five inches tall, and less than one inch thick, the second calculator to use Mostek's calculator-on-a-chip weighed in at a hair over ten ounces, or three hundred grams, when loaded with the four AA batteries that powered it. Its twelve digits, embedded within a silver-and-black case of aluminum and plastic, glowed with the warm red of light-emitting diodes rather than the cool blue-green of the Junior's vacuum fluorescent display.[26] The future had arrived, and boy was it expensive: at a time when the average Japanese graduate earned ¥46,500 a month, the Handy-LE sold for a cool ¥89,900, or $251.[27] The price would not have troubled Aristotle Onassis, ship-

O The Busicom Handy-LE, also known as the LE-120A. This is the world's first
pocket calculator. The switch labeled "4," "2," and "0" controls the number
of decimal places used in calculations. Calculators that provided a "floating"
decimal point were rare. *Häll, Peter. "Miniräknare." Stockholm: Tekniska Museet,
1971. https://digitaltmuseum.se/021026362246/miniraknare. CC BY image
courtesy of Peter Häll / Tekniska Museet.*

ping magnate and second husband of Jackie Kennedy, who is said to
have bought a pile of Handy-LEs as gifts. For everyone else, the built-
in wrist strap helped protect one's investment.[28]

Busicom spun the Handy-LE into a hierarchy of related models.
The LE-120G was plated with 18-karat gold while the LE-120S was
built on a lighter plastic chassis. The LE-100A was a more affordable
ten-digit model, and 1972's LE-80A shrank the Handy-LE into the
same volume as a packet of cigarettes.[29] It was a prescient echo of
today's welter of identikit smartphones.

With the Junior and the Handy-LE, Busicom had proved that
it could streamline its more basic models. But Mostek's calculator-
on-a-chip was a one-trick pony, capable only of basic calculations

in order to keep its price as low as possible. The second prong of Busicom's strategy was quite different. For its high-end, business-oriented machines it would build a "chipset," or collection of integrated chips, that could be reprogrammed to work as any device Busicom might want to sell: a calculator, a cash register, or an automated teller machine.[30]

MASATOSHI SHIMA arrived at Intel's Santa Clara offices on June 20, 1969. The Japanese engineer had come to California to help Intel put eight large-scale circuits of his design into production. Shima's chipset was designed from the start to support a variety of devices: two chips collaborated to perform decimal calculations; a shift register stored numbers in use; and yet other chips interfaced with displays, keyboards, and printers. But it was the inclusion of a chip called a ROM, or read-only memory, that marked Shima's chipset apart from Mostek's calculator-on-a-chip. Shima's calculation chips were unable to do anything much by themselves, but his ROM chip could be custom-wired at the time of manufacture to orchestrate the calculation units' simple operations into more complicated instructions. Instructions that could, for example, run a calculator or an ATM.[31] By reconfiguring this one component, the entire system could be repurposed to create new calculator models or to add new features to existing ones.

How it was that Shima had come up with this design is not entirely clear. One story credits Shima's supervisor, Tadashi Tanba, with having borrowed the concept of a programmable ROM from his days in Japan's burgeoning computer industry. Shima himself said that he had designed his chipset at Tanba's direction while cloistered in Busicom's Osaka office, and that he was unaware of any other efforts to do the same.[32] In an interview conducted in 1994, however, Tadashi Sasaki, Sharp's calculator wunderkind, offered a different story.

Sasaki explained that in 1968, he had led a series of brainstorming sessions to divine the future of calculator chips. It was already possible to buy a calculator powered by as few as four integrated chips, and a move to two- and even one-chip systems (such as the one in the offing as Mostek) was surely only a matter of time. What would come after that? he asked. Most of his engineers imagined that calculator chips would simply become more complex, and more capable, over time. Perhaps a square-root feature or trigonometric functions could be added as transistor counts rose. The lone dissenting voice came from a woman,* a software researcher named Murakami.[33] What if, she asked, we take the constituent parts of a calculator chip—its arithmetic unit, instruction set, memory, keyboard and display interfaces, and so on—and break them out onto individual chips? The same chips, in different combinations, could be used equally well to build a cash register, a calculator, or even a computer, all without the need to design yet more custom chips. It was a quietly revolutionary idea, quite at odds with chip makers' instinct to produce ever-more specialized chips, and it was summarily dismissed by the men in the room. In accordance with traditional Japanese business etiquette, Sasaki followed the majority view.[34]

Yet Sasaki could not shake the feeling that Murakami had been right. Why build chips tied forever to their initial circuit designs when one could reprogram and rearrange a collection of general-purpose building blocks into whatever configuration one needed? Accordingly, he says, he "gave" Murakami's chipset concept to Busicom and Intel to kick-start their new business relationship. Not only

---

* Asked to identify this lone female engineer during a lengthy oral history interview, Sasaki's interpreter explained, "He forgot. She's already married and changed her name." It was only later that Murakami's surname emerged. Her first name is still unknown.

that, but he also claimed to have funneled around ¥40 million, or around $111,000, of Sharp's money to Busicom to fund the development of the new chips.[35]

Did Tadashi Tanba come up with the idea of a general-purpose chipset? Or Masatoshi Shima? Or the unjustly ignored Ms. Murakami? Rather than try to unpick the precise sequence of events from a distance of half a century, it may be enough, as computer historian Ken Shirriff puts it, to acknowledge that this was simply an idea whose time had come—a natural destination for an industry always hunting for the next leap forward.[36]

SHIMA'S VISIT TO INTEL did not start well. Meeting with an engineer named Marcian E. "Ted" Hoff, Shima tried to explain the hows and whys of his design. But Hoff had little experience with chips as complex as those that the Japanese engineer wanted, and Shima had to tutor the American in the intricacies of calculator keyboards, printers, and magnetic card readers.[37] Eventually, Hoff understood Busicom's design, but he did not like it. For one thing, even if Intel had enough staff to turn it into a reality (which it did not), it would have been impossible to bring the chips to production for the agreed sum of a hundred thousand dollars. For another, the finished chipset would be far more expensive than the $50 per unit that Busicom could afford to pay.[38]

Hoff pitched Robert Noyce an alternative approach. Shima's decimal-calculation circuits were too complex and too specialized, he said, and in their stead he proposed a single processing chip that would support only the most basic binary arithmetic—addition, subtraction, doubling, and halving.[39] As with Shima's design, a ROM would arrange these simple operations into more useful ones, and, by the same token, could be programmed to tell the processor how to convert from binary to decimal. Hoff's design would need a memory chip to serve as a scratchpad for the simple-minded processor, but that was fine; Intel had been founded to make memory chips, and it would not be a stretch for Intel's engineers to design a new one. Finally, Hoff asked, why must we have a custom chip for each kind of external device? Shima had imagined that each kind of peripheral device with which his chipset might interact would need a dedicated chip of its own: one for keyboards, one for card readers, one for display circuits, and so on. Hoff proposed to modify a "shift register" chip so that it could act as a universal interface with the world outside the chipset. A shift register to be driven, of course, by instructions baked into the ROM and executed by the processor.[40]

Hoff's design was audacious. Most general-purpose computers of the era had their brains and memories diffused across hundreds or thousands of chips and other components, and the idea that those scattered elements could be replaced by a mere handful of integrated circuits was almost heretical. Granted, Busicom was building a calculator rather than a computer, but Hoff's chipset could theoretically be turned to any computational purpose—decimal math; binary math; cryptography; financial management; any repetitive logical task one might think of—and Noyce was duly convinced.[41] And at a meeting between the companies in September 1969, Shima's boss, Tadashi Tanba, was won over too. Shima put aside his own design and started figuring out how to make Hoff's chipset work as a calculator.[42]

THE SECOND PHASE OF Busicom's collaboration with Intel unfolded about as smoothly as the first. Legal delays set back the project by months, and when Shima returned to California in April 1970 he was enraged to find that little progress had been made.[43] Intel had pinned their hopes on one Federico Faggin, an

Italian engineer at Fairchild who had made several improvements to that company's chip-production process. Faggin had been mystified at the ongoing defections from the well-established Fairchild to Intel, still a precarious start-up, but his curiosity overcame him when he was offered a job there. On April 3, 1970, Faggin's first day at Intel, he was assigned to "the Busicom project"; on his second day, Masatoshi Shima arrived from Japan and demanded to know how Faggin would deliver the chips.[44] As the Italian recalled in 2009, the whole thing seemed impossible:

> I had less than six months to design four chips, one of which
> [the processing unit] was at the boundary of what was possible; a chip of that complexity had never been done before. I had
> nobody working for me to share the workload; Intel had never
> done random-logic custom chips [processors] before, and,
> unlike other companies in that business, had no methodology
> and no design tools for speedy and error-free design.[45]

Nevertheless, Faggin placated Shima with a promise to deliver the first working chips by the end of the year and privately resigned himself to working eighty-hour weeks until then. (Abandoned at home, Faggin's wife, Elvia, returned to Italy with the couple's three-month-old baby to stay with her family.)[46] Faggin and two technical drawing specialists, Rod Sayre and Julie Hendricks, worked ceaselessly to draft the "masks" that would selectively add impurities to the silicon to make transistors and then connect them together using Robert Noyce's patented circuit deposition process.[47] To the mask of the central processing unit, or CPU, Faggin added the Intel logo, the chip's part number (4004, to match the 4001 ROM, the 4002 memory chip, and 4003 shift register), and, triumphantly, his own initials: "F.F."[48]

Somehow, it all came together. In March 1971, after more than a year of false starts, recriminations, and late nights, Intel delivered the first working chips to Busicom. Faggin's 4004 formed the center-

O The "masks," or circuit layouts, for Intel's seminal 4004 CPU. At the top right of the mask, the initials "F.F." declare that the chip was laid out by one Federico Faggin. *Copyright © Intel Corporation.*

piece of the MCS-4, or Micro Computer Set, comprising 2,300 transistors connected by metal pathways ten times thinner than a human hair.[49] Shima wired a handful of the chips into a prototype calculator and, the following month, declared the design to be complete.[50]

Amid all this excitement the calculator itself was almost an afterthought. Busicom called it the 141-PF, and, launched in October 1971, it was mundane in the extreme. It was, for instance, a desktop calculator in a world of increasingly tiny pocket-sized models. Like Canon's Pocketronic, it printed its results at a time when glowing LEDs and fluorescent vacuum displays were all the rage. In appearance it was square and beige, which was a useful metaphor for the level of excitement it might be expected to generate in a potential buyer.[51]

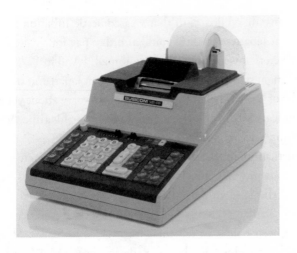

O The Busicom 141-PF, a printing desk calculator whose beige exterior hid a collection of advanced microchips. Among them was Intel's 4004, the world's first successful integrated central processing unit. *Copyright © Intel Corporation.*

Yet on the inside, the 141-PF was revolutionary in a way not seen since the Casio 14-A had first put computer technology in a calculator-shaped box. At its heart was the first successful single-chip CPU: two thousand transistors; two hundredths of a square inch of doped silicon; a brain in miniature; a harbinger of a future not yet written. The *Wall Street Journal* would call the 4004 "the chip that changed the world," and it could be yours for ¥159,800, excluding shipping.[52] It even came with a free calculator.

ULTIMATELY, neither Intel's 4004 nor Mostek's calculator chip could save Busicom. Protectionism aside, Japan's economic trump card had always been its supply of cheap labor, but, ironically, that advantage was eroded by smaller, simpler chipsets that were easier to install.[53] Piled onto that was domestic inflation that drove down what Japanese workers could afford to pay for a calculator and a strong yen that was doing the same abroad.[54] And in 1972, with the establishment of diplomatic relations between Japan and the People's Republic of China, Japanese firms began to worry about yet another geographical shift in the electronics industry—this time, across the East China Sea.[55]

American manufacturers smelled blood in the water. If a Japan down on its luck was not inducement enough, there was the small matter of the $241 million that U.S. buyers were estimated to have spent on calculators in 1970 alone.[56] Where once American companies had been happy to sell their microchips to Sanyo, Sharp, and Sony, or to rebadge calculators made by those same Japanese corporations, now the shelves of Sears and JCPenney were increasingly stocked with honest-to-goodness, made-in-the-USA models. Old brands mixed with new ones so that one could buy a calculator made by Bowmar, Burroughs, Commodore, Monroe, Rockwell, Singer, Smith Corona, or Victor and be happy that one's dollars were staying local—unless, of course, one happened to pick one of the many models still outsourced to a Japanese frenemy.[57]

One player in particular had learned a hard lesson on that front. Two years after the damp squib of Canon's Pocketronic, and with its yield problems finally under control, Texas Instruments debuted its first U.S.-made, own-brand calculator in the summer of 1972.[58] The TI-2500, or "Datamath," was the first of a calculator dynasty: by 1975, TI had become the largest calculator brand in the United States; by 1980, it was one of the *only* calculator brands in the United States.[59]

Busicom never lived to witness TI's rise to the top. Beset by plummeting calculator prices, execs at the Japanese corporation went to Intel just a month after the 141-PF had gone on sale to plead for cheaper chipsets. Ted Hoff and Federico Faggin counseled Robert Noyce to agree, but only if Intel was allowed to sell MCS chips, then exclusive to Busicom, to other customers. Many of Intel's senior managers did not want the distraction of another kind of chip to sell, however innovative it might be, but Noyce was persuaded, and that was enough. [60] Intel gave back $60,000 of Busicom's money in exchange for the rights to sell the MCS-4 to anyone except another calculator manufacturer, and the first adverts for the chipset ("a new era for integrated electronics!") appeared in November 1971. [61] The rest was history.

Shortly thereafter, Busicom was history too. Having developed the first real pocket calculator and pioneered the modern CPU, the company lost its footing during the ferocious calculator price wars of the early 1970s. By 1974 Busicom was bankrupt, its name bought by a small British office-machine business and the Japanese mothership dismantled for parts. [62] From now on, only the very smart, the very large, and the very lucky would stand a chance in the cutthroat world of the pocket calculator.

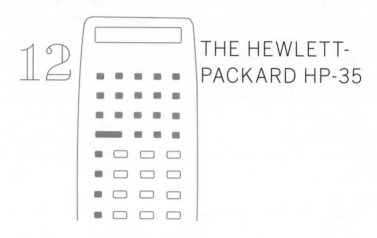

12 THE HEWLETT-PACKARD HP-35

CALCULATORS HAD EVOLVED from the tube-filled TARDIS-like console of the Sumlock Anita to the pocketable minimalism of the Busicom Handy-LE in less than a decade. But little had changed in what those calculators could do: the holy quaternity of add, subtract, multiply, and divide were all that could be expected from the average electronic calculator. For those who needed more serious math, a pen, paper, and a slide rule were nonnegotiable.

Until, that is, a hitherto obscure electronics manufacturer named Hewlett-Packard exploded the status quo with a million-dollar rush job to build the best pocket calculator the world had ever seen. The remarkable thing about that calculator, the HP-35, was that it did not rely on some breakthrough in math, material science, or manufacturing. There was no newly invented transistor, silicon chip, or CPU behind its success. Instead, there was something more prosaic: plain old brain power, as embodied by the algorithms encoded in its electronics and the careful rigor of its design. The HP-35 ran on canned intellect.

T HE CARPENTER FINISHED his work. The gaggle of engineers who had gathered in the office of their absent boss, Bill Hewlett, pushed the sliding typewriter pedestal of Hewlett's L-shaped desk
back into its slot. It fit perfectly, as it had always done, but now the
sleek gray desktop calculator perched on top of it fit, too, passing neatly
through the slot the carpenter had just enlarged by an eighth of an inch.[1]

Bill Hewlett was the president of Hewlett-Packard, the electronics business he had co-founded in a Palo Alto garage in 1939 and that
by now, in 1968, employed thirteen thousand people and pulled in
a quarter of a billion dollars per year.[2] The calculator was the company's first, the HP 9100A, and Hewlett had insisted that it had to
fit within his desk's typewriter compartment. When the engineers

O A prototype of the HP 9100A, Hewlett-Packard's first calculator, whose circuitry
and algorithms would inform the development of the handheld HP-35. *"HP9100
Prototype Desktop Electronic Calculator." National Museum of American History.
Washington, DC: Smithsonian Institution, 1966. https://americanhistory.si.edu/
collections/search/object/nmah_1196393. Image courtesy of Division of Medicine
and Science, National Museum of American History, Smithsonian Institution.*

working on the project found out it would *not* fit, they waited until Hewlett was on vacation and arranged for some clandestine carpentry.[3] "You see!" said Hewlett on his return, "I knew you could do it!"[4]

Like Olivetti's Programma 101 of 1965, of similar dimensions and from which HP had drawn inspiration of a legally consequential magnitude, the 9100A was programmed via magnetic cards and could, by means of an optional printer, create a hard copy of its calculations. Unlike the P101, the 9100A had a built-in cathode ray tube, like a miniature television, that displayed the contents of the calculator's short-term memory, or "stack." Moreover, the 9100A offered a host of advanced functions—sines, cosines, logs, exponentials—that were more usually found in books of mathematical tables.[5] It was a prince among calculators, and Bill Hewlett wanted one in his pocket. "About fifteen minutes or so after we finished the 9100," one engineer recalled, "Bill Hewlett said we should have one in a tenth the volume, ten times as fast, and at a tenth the price."[6] Hewlett later refined his demands to a scientific calculator (that is, one that could perform the 9100A's advanced mathematical functions in addition to the more usual addition, subtraction, multiplication, and division) that would cost less than $500, weigh less than nine ounces, and fit in a shirt pocket.[7]

This was a bold target, given that in 1968 the electronic pocket calculator did not yet exist in any meaningful way. Texas Instruments had completed their Cal Tech prototype in 1966 but could not make it cheaply enough to sell as a product. Across the Pacific, Sharp, Sanyo, and others were readying handheld calculators based on integrated chips, but these were chunky, simplistic machines capable only of the four basic arithmetical operations.[8] Nor was it clear that anyone would buy the sort of calculator that Hewlett wanted to build, even if his engineers succeeded in building it. One of HP's own marketing studies found that the typewriter-sized, $5,000 9100A occupied the sweet spot for scientific calculators.[9]

And yet, Hewlett's goal of a pocket-sized 9100A was not as wild as it might have seemed. To start with, Japan's focus on cheaper models left an obvious gap above them for more expensive calculators.[10] And the horizons of microchip technology were constantly expanding: Gordon Moore, who would later co-found Intel, published an article in 1965 in which he predicted a doubling of chip complexity every year.[11] What we now call Moore's Law meant that a chip that was impossible in 1968 might be within reach a year or two later. Hewlett was betting that it would be possible to shrink down the 9100A sooner than his competitors could enhance their simpler machines.

WITH HP TO HIMSELF (Hewlett's founding partner, Dave Packard, was moonlighting as Richard Nixon's deputy secretary of defense), Hewlett asked his industrial design department to imagine how a pocket 9100A might look.[12] It was a challenging brief. HP's catalogues were stuffed with oscilloscopes, audio-signal generators, and other gadgets whose boxy, unlovely casings had been rudely wrapped around preexisting internal components. This would be the first time that HP had tackled the outside of a device before its insides—a first, perhaps, for the electronics industry as a whole.[13] Nor was there anywhere to crib ideas from. There was no other calculator like the one in Bill Hewlett's head, and real-life pocketable calculators were still a year or more away.[14]

Edward Liljenwall, a longtime HP designer, was given the job. He knew that the calculator would need thirty or more keys, that it would be powered by a handful of AA batteries, and that it would have to display between twelve and fifteen red LED digits. Most other details were hazy. Liljenwall plowed ahead, drawing a paperback-sized case that sloped toward the user and whose angled sides made it seem smaller than it really was. Its display was canted so that it could be seen both on the desk and in the hand, and its color-coded buttons, arrayed below the display, were divided among numeric, arithmetic, and scientific functions.[15]

Those keys had been especially tricky. They were necessarily smaller and closer together than those of a desktop calculator, but after fiddling with their size and layout, Liljenwall arrived at what seemed to be a usable arrangement.[16] Absent any comparable devices, the accordion had provided an unlikely touchstone: if an accordionist's fingers could dance across hundreds of small, tightly packed keys, it was reasonable to think that an engineer or mathematician would have no trouble with the buttons on Liljenwall's pocket calculator.[17] (Later, some users of the new calculator would operate it with their thumbs, like a smartphone.)[18]

O   A watershed moment in calculator design: the HP-35 married sophisticated features to a subtle, clever design. Unprecedented sales put Hewlett-Packard on a path from the lab to the living room. *Scientific Apparatus And Instruments.* "Hewlett-Packard Model 35 Calculator with Case." Philadelphia, PA: Science History Institute. Accessed June 7, 2022. https://digital.sciencehistory.org/works/jd472x57t. Public domain image courtesy of the Science History Institute.

○———○

B Y 1970, with a product design in hand, Bill Hewlett began pressing an engineer named Dave Cochran to decide how the calculator's internals would work. Cochran had been the architect of the 9100A's electronic brain, and he was the natural choice to repeat the job for its pocket-sized sibling.[19] But where the desktop model had relied on discrete components and printed circuit boards, the only way to fit those same circuits into Liljenwall's svelte enclosure was to embrace the integrated chip.[20] Late in the year, Cochran presented a plan detailing the chips he would need and how they would work together, and he asked Hewlett for the million dollars he figured it would cost to develop them.[21]

For the first time, Hewlett balked. Cochran protested: "It's what you've been asking for two years," but the United States had been in a recession since 1969, and a cash-conscious HP was sending its workers home every other Friday without pay.[22] Hewlett told Cochran he'd commission a study to find out how well his pet project might sell and left the engineer hanging.

That might have been the end of it, except that Cochran and a collection of fired-up colleagues kept working on the calculator regardless, surreptitiously wedging in budget requests wherever they would fit. The results of Hewlett's marketing study, when they came back, were helpfully inconclusive. No one knew how to put a price on a scientific pocket calculator or, indeed, whether anyone would buy it. Hewlett knew only that *he* still wanted the calculator and so he gave the go-ahead, budget be damned.[23] Founded on Hewlett's zeal, funded by Cochran's asked-for million, and executed on a strict schedule by employees poached as needed from across HP's divisions, the pocket calculator project became a moonshot for the normally conservative company.[24]

○———○

C OCHRAN'S DESIGN called for five chips of three different types: an arithmetic chip capable of simple decimal arithmetic; three read-only memory chips, or ROMs; and a control unit to run the show.[25] It was, essentially, a halfway house between Busicom and Intel's respective designs for the 141-PF: simpler than the multitudinous special-purpose chips of Masatoshi Shima's original proposal, but more specialized than Ted Hoff's general-purpose response to it. Totaling around thirty thousand transistors shared among its five chips, HP's chipset was perhaps the most sophisticated set of integrated circuits then in production.[26] The new calculator would need every ounce of that sophistication.

Echoing the Intel 4004, the arithmetic unit of HP's pocket calculator was exceptionally simple, supporting only addition, subtraction, and, because it operated on decimal numbers rather than binary, multiplication and division by ten rather than by two.[27]

This ability to multiply and divide only by a fixed amount is called "shifting," and it is a quirk of the way that computers store numbers. (Shift registers, for instance, can both store data and "shift" it in exactly this way.) The Intel 4004 worked on base-2, or binary numbers, and so shifting a number either multiplied or divided it by 2; HP's calculator ran on base-10, or decimal numbers, and so shifted numbers were multiplied or divided by 10.

With only this basic palette of operations at his disposal, Cochran would have to program the calculator's ROMs to perform math ranging from decimal multiplication all the way up to logarithms and trigonometric functions. To do so, he created a library of *algorithms*, or numerical recipes, that could decompose a square root, a cosine, or other operation into a sequence of additions, subtractions, and shifts.[28]

Though, as we've seen, the word "algorithm" comes from the name of the ninth-century mathematician al-Khwārizmī, the concept is much older. The ancient Egyptians, for example, invented a method, now called *duplation*, with which to multiply large numbers using only the 2 times table and a few careful additions.[29] The Babylonians knew how to approximate square roots using a method later expanded on by Isaac Newton.[30] And in the sixteenth century, John Napier constructed an algorithm for calculating logarithms, which in turn could be used in yet another algorithm to simplify difficult multiplications. Even the long division and grid multiplication methods taught at elementary schools are algorithms in their own right.

It was in the twentieth century that algorithms and computers truly came together. In 1928, a German mathematician named David Hilbert posed a question, the *Entscheidungsproblem*, that asked whether it was possible to construct an algorithm that could decide whether a given mathematical statement was true or false. (Notably, Hilbert did not ask his readers to come up with such an algorithm, only to say whether it was possible to create one.) It was a difficult problem to grasp, much less to answer, but it was taken up with enthusiasm by a young Alan Turing.[31]

To address Hilbert's conundrum, Turing first had to define what an algorithm was. He settled for a definition that hinged on the idea of *computable* numbers, or numbers that can be stored by a machine, and devised a simple imaginary computer that was theoretically capable of executing all conceivable mathematical algorithms.[32] He called it a *universal machine*. It was the first formal definition of a general-purpose computer, and it had arisen out of a question that spoke to the very nature of mathematics. And if, as any physicist will tell you, the existence of mathematics is an inevitable consequence of the existence of the universe, then the invention of the computer is something that any halfway sapient life form will eventually aspire to.[33] To exist is to compute, and to compute is to embrace the algorithm.

All this is to say that by the time that Dave Cochran had to squeeze the brains of a scientific calculator onto a few slivers of silicon, a *lot* of thinking had been done on the subject of algorithms. A lot rode on it, too, since the behavior of a computer or calculator depended as much on the algorithms in its ROM as on the design of its CPU. As a case in point, Bill Hewlett had once demonstrated the 9100A to An Wang, the head of a competing firm, who came away both impressed and daunted by the speed of the HP machine. Wang later protested that HP must have stolen a particular algorithm used in his own machines, but Cochran knew otherwise. He sent a note suggesting that Wang should consult Henry Briggs's 1624 book *Arithmetica logarithmica* to see that John Napier's bromantic acquaintance had invented that same algorithm three centuries before the quarrelling Americans.[34]

WHEN IT CAME TO implementing the math within the new calculator, Cochran's choice of algorithms was an exercise in contrasts. To multiply and divide, he used a so-called shift-and-add algorithm derived from the ancient Egyptian method of duplation; for almost everything else, he used a method invented just a decade earlier for use in a jet-powered nuclear bomber.[35] Either one would serve to illustrate how algorithms can turn simple calculators into much more capable ones, but only the first can be explained in a paragraph rather than a chapter. For the sake of brevity, then, shift-and-add it is.

Using only the calculator's addition, subtraction, and multiply/divide-by-ten features, multiplication was carried out something like this:

1. Let x be the first number, or multiplicand
2. Let y be the second number, or multiplier
3. Let t be the total, or product. Set this to 0.
4. Repeat until the last digit of y is 0:
   a. Add x to t
   b. Subtract 1 from y

5. If y is 0, end the calculation. Otherwise,
   a. Shift x by one position to multiply by 10
   b. Shift y by one position to divide by 10
   c. Repeat from step 4

This definition has a number of weaknesses, such as the handling of fractional numbers and those with different signs. Yet even this simplified example shows how a carefully crafted algorithm can work with and around the limitations of a given chipset.

To see this shift-and-add algorithm in action, consider multiplying 12 by 13. We go through a series of steps as follows:

Step	x	y	t	Notes
1, 2, 3	12	13	0	Initial set-up
4	12	$13 - 1 = 12$	$0 + 12$ $= 12$	The last digit of y is not 0, so add x to t and subtract 1 from y.
4	12	$12 - 1 = 11$	$12 + 12$ $= 24$	Repeat previous step.
4	12	$11 - 1 = 10$	$12 + 24$ $= 36$	Repeat previous step.
5	$12 \times 10 = 120$	$\dfrac{10}{10} = 1$	36	The last digit of y is now 0, so multiply x by 10 and divide y by 10. The product is unchanged.
4	120	$1 - 1 = 0$	$36 + 120$ $= 156$	The last digit of y is not 0, so add x to t and subtract 1 from y.
5	120	0	156	y is 0, so the multiplication is complete.

That, in a nutshell, is it: an algorithm is nothing more than a way to guide a computer (whether human or electronic) through a more complex calculation. Cochran used a modified version of the same method to handle division, but the lion's share of the new calculator's functions were provided by a much more sophisticated algorithm.

CORDIC, or "COordinate Rotation DIgital Computer," was a product of the Cold War. In June 1956, Jack E. Volder of the Convair aircraft company was part of a team seeking to improve the navigational computer of the B-58 Hustler, a delta-winged bomber capable of hauling a nuclear weapon more than two and a half thousand miles into the Eastern Bloc but which relied on inaccurate, analog devices called "resolvers" to find its way there and back.[36] Volder had discovered a way to replace these unhelpful devices: a trigonometric algorithm that used only addition, subtraction, and shift operations, and which, as such, could be embedded in the kind of compact digital computer that might fit into the Hustler's cramped fuselage.[37]

CORDIC got short shrift from the U.S. Air Force, which was more interested in hardening computers against atomic blasts than it was in advances in algorithmic science. Eventually, however, Volder was given permission to make CORDIC public.[38] He announced it at a computing conference in 1959, by which time he had worked out how to extend the algorithm to logarithms, exponentials, multiplication and division, and even conversion between binary and decimal numbers.[39]

By 1965, Volder had left Convair and, with a physicist named Malcolm Macmillan, was preaching the CORDIC gospel to whoever would listen. It was in that year that Macmillan visited Hewlett-Packard, bent on selling them a desktop computer powered by Volder's algorithm, but the pair's unreliable machine failed to impress. Later, however, as HP engineers pondered how to cram a computer's worth of logic onto Bill Hewlett's desk, CORDIC was the answer. And it was even more relevant a few years later when Dave Cochran was tasked with downsizing the 9100A to the size of a cigar case.[40] COR-

DIC was the magic dust that would turn Bill Hewlett's calculator from an obsession into a reality.

BY 1972, the calculator was ready. Christened the HP-35 for its thirty-five keys, the finished machine was an engineering triumph.[41] It fit into Bill Hewlett's shirt pocket as requested, even if, as Dave Cochran later quipped, "we thought for a while we'd have to get hold of his tailor."[42] It could reckon with numbers as small as 0.000000000000000000000000000000000000000000000000000 0000000000000000000000000000000000000000000000000000 01 (that is, a one preceded by ninety-nine zeroes) or as large as 100 0000000000000000000000000000000000000000000000000000 00000000000000000000000000000000000000000000000000 (a one followed by ninety-nine zeroes). It was precise enough to work with most of the esoteric numeric values that govern the known universe—G for gravity, c for the speed of light, h for Planck's constant; and more.[43] It could calculate a logarithm, sine, or exponential in less than a second.[44] And, as discovered during a demonstration by a clumsy Bill Hewlett, it could survive a three-foot drop onto a concrete surface, a test that would henceforth be administered to all new HP calculators.[45]

At the cost of a million dollars and forty person-years, HP had liberated the power of scientific calculation from the office desk and put it into the hands (and the pockets) of those who needed it most.[46] But selling the HP-35 was almost as daunting as building it in the first place. There was no question that the HP-35 would be sold to the public: profit margins would be slim, even on a planned production run of 100,000 units sold at $395 each (around $2,600 in today's money), and so selling the calculator via HP's network of commission-hungry sales representatives was never going to work.[47]

Instead, the HP-35 was promoted with sultry photographs of female models, advertised in *Scientific American* and *Esquire*, and sold at Macy's, as if it were a household appliance or a bottle of aftershave rather than a scientific instrument.[48] And sell it did.

Engineers bought HP-35s with their own money, rather than wait for their employers to process purchase orders.[49] College students sold their cars to fund HP-35s. General Electric asked to buy twenty thousand units.[50] Scientists working for the U.S. Army invented a training course* as an excuse to buy HP-35s as "training tools."[51] Steve Wozniak, then an HP employee and later the co-founder of Apple Computer, was hypnotized by the sleek, pocketable machine.[52] And the HP-35 breached frontiers both on Earth and in heaven: climbers took them to the top of Mount Everest; TV broadcast engineers brought them on Richard Nixon's 1972 trade mission to China; and NASA astronauts flew with them to the Skylab† space station.[53] In its first year, HP sold ten thousand HP-35s each month against initial predictions of just a tenth that figure, and would likely have sold more had there been any spare capacity to make them. Profits from the calculator almost singlehandedly pulled HP out of its ongoing slump.[54]

In short, the HP-35 was a roaring commercial success. But Bill Hewlett's pocket marvel was more than just a moneymaker. First, it dealt a hammer blow to the centuries-old slide rule. The four-function pocket calculators already creeping onto the shelves had put the slide rule on notice; now, the HP-35 made it look shockingly

---

* That training was very much needed, since the HP-35 used a concise but tricky style of input called reverse Polish notation, or RPN. Some scholars argue that RPN's very inaccessibility gave the HP-35 an air of exclusivity—yet another reason to stump up $400 for one.

† The Skylab astronauts' ground-based colleagues were also fans of HP's new calculator. So many HP-35s were stolen at NASA that engineers had to lock their calculators to their desks.

anachronistic. Slide rule production lines were closed and inventories were sold off. It was the end of an era.[55]

Second, the frenzy for HP's new calculator was the first instance of a phenomenon that has since become part of the consumer electronics firmament. For all the advances in electronics over the preceding two decades, the visible results had been evolutionary rather than revolutionary: transistorized radios were still radios; transistorized televisions were still televisions; transistorized calculators were still calculators. The HP-35, by contrast, with its incredible precision and its scientific functionality, improved the state of the art in pocket calculators so comprehensively that it seemed to come straight out of *Star Trek*. It not only eclipsed what had gone before but also cracked open a whole new market at the same time, anticipating the manias that would accompany the digital wristwatch, the cellphone, and the personal computer. The HP-35 was the First Big Thing, and it made us perpetually hungry for the next.[56]

13 THE PULSAR TIME COMPUTER CALCULATOR

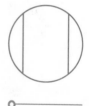N MAY 6, 1970, Johnny Carson's nightly television show opened not with a guest but a gadget. After announcing the evening's roster—entertainer Jack Benny, singer Roberta Peters, actor René Auberjonois—Carson held up a plain metal wristwatch with a dark, blank lozenge at the center of its face. As he shielded the watch from the bright studio lights, Carson asked his co-host, Ed McMahon: "You've seen a digital alarm clock, right?" With that, Carson pressed the watch's crown to illuminate its dark screen with the time in glowing red numerals. "That's wild. That's gonna sell," McMahon responded. "A machine finally put Mickey Mouse out of work," quipped Carson, referencing the line of popular wristwatches adorned with Walt Disney's most famous cartoon character.[1]

The watch that Carson had introduced was the first of its kind. Other than buttons to set or show the time, the Pulsar Time Computer had no moving parts: it was regulated by a vibrating quartz crystal, driven by an integrated chip, and surmounted by light-emitting diodes that displayed the time on demand.[2] Despite its whiz-bang name, it was not a computer. It was not even a calculator.

But the Pulsar Time Computer would soon be united with one in an act that hinted that the pocket calculator juggernaut might finally be running out of road.

THIS HISTORY OF the wristwatch, like that of the pocket calculator, stretches back thousands of years. The sundials and water clocks of the ancient world were joined toward the end of the thirteenth century by mechanical clocks driven by falling weights.[3] In the fifteenth century, portable, spring-wound clocks arrived on the scene; and in the seventeenth century, more accurate pendulum clocks made their earliest appearances.[4] The first watches small enough to be carried on the person arrived in the sixteenth century, and finally, in the nineteenth century, women participating in sports such as cycling and archery drove the adoption of the practical, fashionable wristwatch.[5] From that point on, however, technological innovations were modest. With their product now largely perfected, watchmakers were chiefly concerned with churning out more, cheaper watches rather than improving their form or function.[6]

Essentially all mechanical watches made since the seventeenth century have been regulated by oscillating, spring-driven "balance wheels." At each tick of the mechanism, an impossibly fine "hairspring" forces a weighted wheel to rotate forward; then, when the spring has reached its maximum extension, it contracts and pulls the wheel back in the opposite direction. And so on, and so on. The strength of the hairspring and the rotational inertia of the balance wheel are chosen so that the wheel oscillates at a predictable frequency.[7]

All this changed in 1957. That was the year that a Swiss watchmaker called Bulova unveiled a watch that did away with the balance wheel and hairspring in favor of a tiny tuning fork, vibrated by electromagnets, in service of more accurate timekeeping.[8] The idea of using a tuning fork to regulate a clock had been proposed a hundred years earlier and pursued with varying degrees of success since then, but Bulova's "Accutron," which went on sale in 1960, was the first wristwatch to use the technology.[9] Powered by a tiny button-shaped battery, the watch's single transistor helped generate a magnetic field that kept the tuning fork humming at the correct pitch.[10]

Earlier the same year, the Hamilton Watch Company of Lancaster, Pennsylvania, had announced its own battery-powered watch. Hamilton's version kept the balance wheel but drove it with an elaborate arrangement of magnets and electrical contacts.[11] Worried that Bulova would steal its thunder, the American company rushed its "Electric" watch into shops in a variety of futuristic casings. But watchmakers and jewelers had no experience with this intricate new timekeeping movement, and complaints of irreparably broken Hamilton Electrics soured the public against the model and the brand.[12] The Pulsar was to be Hamilton's comeback, but the company was not alone in aiming to break new ground.[13]

I N 1967, Hamilton had commenced and then abandoned an effort to build a watch regulated by a quartz crystal. The phenomenon of piezoelectricity, in which certain crystals generate an electric current when put under physical stress, had been discovered in 1880 by Marie Curie's fiancé, Pierre Curie. (The inverse effect, when a crystal deforms in response to an electric current, was discovered the following year.)[14] It was not until the early 1920s, however, that researchers began to experiment with the use of crystals in place of tuning forks. Connected to a suitable electrical circuit, a quartz crystal could be made to resonate at an incredibly precise frequency, promising to make domestic clocks and watches more accurate than ever before.[15]

The spanner in Hamilton's works was news that both Seiko, a Japanese watchmaker, and a consortium of Swiss companies were close to completing their own quartz-driven watches.[16] Seiko got there first with its "Astron" analog wristwatch of 1969, an understated horological masterwork whose single integrated chip contained seventy-six transistors and whose quartz crystal "ticked" at 8,192 times per second. The Astron's only moving parts were an electric motor and the hands on its face.[17]

Beaten to the finish line yet again, Hamilton needed to try something new. What if, its bosses wondered, a watch could be made with no moving parts at all? After all, Texas Instruments, Sharp, and others had shown that it was possible to build a fully transistorized calculator without a spring or a motor in sight. The requisite technology was already available, or close to it; all that was missing was the ambition to do something with it.

How Hamilton came to set itself this goal is still a mystery, but the seed may have been sown when the company made a handful of futuristic analog watches for Stanley Kubrick's 1968 film, *2001: A Space Odyssey*. Hamilton had also given Kubrick a table clock that used Nixie tubes to display the time and date, but it never made it to the screen. One story is that John Bergey, Hamilton's head of research and development, got the idea for a digital watch from that same table clock. Another says that Bergey was already researching digital watches, but that he held them back from use in the film to throw his competitors off the scent.[18] Either way, Bergey was the driving force behind the digital-watch project: He oversaw its difficult development; he chose its space-age name after reading about astronomical bodies named pulsars; and he was entrusted to carry one of the only three or four prototypes then in existence to the May 6, 1970, taping of *The Tonight Show Starring Johnny Carson*.[19]

The finished product arrived in stores in 1972. It had taken those two extra years to fit the Pulsar's 1,500 electronic components onto a single, affordable chip, and to figure out how to stop battery acid from leaking out and scorching the wearer's skin.[20] (Hamilton's engineers displayed their burned wrists with pride, since few people on the planet at the time could claim to have worn a digital wristwatch.)[21] Then there was the matter of the Pulsar's battery life. Its LED display lit up only when a button was pressed and would have drained the watch's batteries in mere hours otherwise.[22] But nothing more could be done to extend the battery life, and so the solid-gold Pulsar went on sale regardless, priced at an eye-watering $2,100—about the same as a small car, and more even than a gold Rolex.[23]

The Pulsar immediately became a must-have trinket for the rich and famous. Haile Selassie, the last emperor of Egypt, had one, along with the Shah of Iran and the King of Jordan. Sammy Davis Jr., Keith Richards, Elton John, and Roger Moore all sported Pulsars, as did Moore's cinematic alter ego, James Bond. Presidents Richard M. Nixon and his successor, Gerald Ford, were both Pulsar fans,

O   The stainless-steel variant of the original Pulsar watch, as sold in 1972 for $275. *Image courtesy of Igor Tesic (https://20centurywatches.com).*

and U.S. senators vied to be the first to wear one to Congress.[24] A stainless-steel model was released later in 1972, selling at a much thriftier $275, and Hamilton could not keep up with demand.[25] The Pulsar had sparked a digital-watch hysteria.

HAMILTON SOON FOUND itself with competition, but not from the traditional watchmakers over in Switzerland. Riding in Hamilton's wake were companies more often associated with calculators than with timepieces, such as National Semiconductor, Texas Instruments, and Robert Noyce's old company, Fairchild.[26] The few conventional watchmakers who chose not to make LED wristwatches in the Pulsar's image—Seiko, Patek Philippe, and Timex among them—suffered mightily as a result.[27] Yet others were forced to grovel at Hamilton's door, with the Swiss watchmaking stalwart Omega reduced to rebadging American-made Pulsars.[28] "Maybe we'll introduce a calculator for $1.98," growled a watchmaking executive at a trade show in 1975, "and mess up the semiconductor industry's calculator market the way they've messed up our watch market."[29]

Accordingly, Hamilton set about broadening their lineup. A Pulsar that could tell the date as well as the time arrived late in 1973; ladies' models, which squeezed the Pulsar's bulky LEDs, circuitry, and batteries into marginally daintier cases, came the next year. In 1975 came a version whose screen illuminated with a flick of the wrist.[30] But the rush to make, and to profit from, this most desirable type of watch put prices into freefall. In 1973, Hamilton had charged $275 for a stainless-steel Pulsar, but by 1975, the Sears Christmas catalogue could list a watch with identical features for a mere $49.99.[31] Yet again, Hamilton had to innovate to survive—and what better way to do that than to cross over with the decade's *other* hot new electronic product?

The pocket calculator had come a long way since Busicom's Handy-LE of March 1971.[32] Hewlett-Packard's HP-35, unveiled a year later, put advanced mathematical functions into its owners' pockets, and 1974's HP-65 added programmable magnetic cards on top of that.[33] And if the HP-65's asking price of $795 was a little too high for most buyers, there were plenty of other options for much less. Casio's bare-bones "Mini" calculator debuted at less than a hundred dollars in August 1972, and was selling for $59.99 by November that year.[34]

Over in the United Kingdom, an entrepreneurial engineer named Clive Sinclair was making waves with his own line of affordable calculators. Sinclair used off-the-shelf chips that normally required bulky AA batteries, but he had found that the chips held an electrical charge long enough so that he could periodically interrupt their power supply without any ill effects. The Sinclair "Executive" of 1972 took full advantage of this, putting a four-function calculator into a sleek black case a third of an inch thick and powering it with tiny watch batteries.[35] There had been pocket calculators before, but the Executive was the first to truly embody the term, and all at the comparatively modest price of £79 ($193).[36]

This rush of innovation led to a point in the mid-1970s when it was possible to buy a calculator that was small, powerful, and/or stylish in whatever combination one's bank balance would support. For the typical well-heeled watch aficionado, of course, money was not such a pressing concern. And so, as Hamilton looked to revitalize both its flagship product and its flagging profits, the answer must have been obvious: lash a digital watch to a calculator, slap on an exorbitant price tag, and watch the cash roll in.

○——○

T O WATCH-COLLECTING WONKS, the story of the Pulsar's startling rise and (look away now) equally sudden fall is as fascinating a subject as celebrity timepieces such as Buzz Aldrin's moon-bound Omega or Paul Newman's $18 million Rolex.[37] As such, there have been many hagiographies of the Pulsar Time Computer. Curiously, however, very little has been written about its cumbersomely named successor, the Pulsar Time Computer Calculator.

The basic facts are simple enough. Just before Christmas in 1975, Time Computer, Inc., Hamilton's newly spun-out digital watch subsidiary, launched a version of the Pulsar with an integrated calculator.[38] Only a hundred examples were made at first (more were to come the following year), with each one encased in 18-karat gold and priced at a cool $3,950. (Gerald Ford asked his wife for one as a Christmas gift. On learning the price, she demurred.)[39] The lucky few who managed to snag an example got a squared-off sledgehammer of a watch studded with twenty-two tiny buttons, a stylus with which to tap them, and whose LED display, as ever, remained dark until called forth with the press of a key. In calculator mode, invoked by the "0" key, the Pulsar could add, subtract, multiply, divide, and compute percentages up to twelve digits, even if only six at a time would fit onto its tiny screen.[40]

Internally, the Time Computer Calculator looked like exactly what it was: a watch and a calculator cohabiting the smallest possible space. It was powered by a handful of custom chips designed by Time Computer, Inc., and manufactured by, among others, Japan's Toshiba corporation, between which the watch's timekeeping and calculator functions were distributed. The chips, just millimeters across, did without the multilegged ceramic packages in which integrated circuits were normally encased and were soldered directly onto a printed circuit board.[41] Such was the watch's complexity that its service manual warned jewelers not to attempt any repairs: the only way to fix a broken Time Computer Calculator was to replace its guts entirely.[42]

O The Pulsar Time Computer Calculator: the first calculator watch, complete with combination pen and stylus with which to tap its buttons. *Image courtesy of Igor Tesic (https://20centurywatches.com).*

In keeping with the original Pulsar, a steel variant of the calculator watch went on sale in 1976, accompanied by a drop in price from $3,950 to $550.[43] This time, however, the Pulsar's grace period was even shorter, and the same year saw the arrival of a pair of competitors. Incongruously, the first of these came from the microelectronics division of Howard Hughes's aircraft company, which, like Texas Instruments, had lately found its military contracts on the wane. Hughes had seized on the digital-watch craze as a means to occupy its idle chipmaking facilities and became an OEM manufacturer for several different brands. Its calculator watch, a virtual clone of the Pulsar, was sold variously as the "Compuchron" and the "Timeulator."[44] Hewlett-Packard's HP-01, by contrast, was a far more considered effort. HP's watch could calculate with time and date measurements, handled the same wide range of numbers as the HP-35, and could multiply or divide stopwatch timings by a fixed value—handy for tracking the cost of expensive long-distance phone calls, for example.[45]

○———○

T HE POCKET CALCULATOR, all the while, had been going gangbusters. Production had almost doubled every year since the introduction of the first handheld calculators, so that Japan alone ramped up from 1.5 million per year in 1970 to 10 million in 1973.[46] Worldwide, that figure went from 14 million in 1973 to 50 million two years later.[47] And prices had dropped even faster than production had risen: in 1973, the UK's *New Scientist* magazine estimated that the cheapest calculators were halving in cost every six months.[48] By Christmas 1974, it was possible to buy an electronic calculator for only £10, or about $23.[49] By the middle of the decade, however, in the wake of a series of economic shocks, the calculator boom had run out of steam.

In 1971, to resuscitate the flagging U.S. economy and avoid a looming run on gold, Richard Nixon had untethered the U.S. dollar from the gold standard. Then, in 1972, cereal harvests failed in several major grain-exporting nations. And in 1973, the oil exporters of the Arab world announced an oil embargo in retaliation for U.S. support for Israel during the Yom Kippur War. For the United States, the result was a deep recession that troughed in 1975.[50] A news bulletin in the industry journal *Electronics* wrote glumly of "slumping calculator sales."[51] And amid weakening demand for its products, Rockwell, a giant of the chipmaking industry, began to wean itself off the calculator chips that made up half of its electronics business—and this from a company that made more than 30 percent of the world's calculator chips.[52]

The slowdown prompted calculator makers to change direction. There was still room for ever-smaller, ever-slicker models, certainly, but now manufacturers were asking themselves a new question: Who does not already have a calculator, and how can we get them to buy one? If the Pulsar Time Computer Calculator was one answer, there were plenty of others.

The quickest way to broaden the appeal of one's products was to put an otherwise normal calculator into a novelty case. A German brand named Triumph-Adler showed the way with its matching "Sir" and "Lady" calculators of 1975, finished respectively in serious silver and black and coquettish red and gold.[*] Standard features included eight-digit vacuum fluorescent displays, arithmetic operations and lazy gender stereotyping.[53] For a more American take, look no further than Disney's inevitable "Mickey Math," a simple four-function calculator encased in robust, child-friendly plastic and emblazoned with Mickey Mouse in wizard garb.[54] Or, indeed, Texas Instruments' "Spirit of '76," a budget-priced TI-2000 dressed up in fancy red, white, and blue duds for the USA's bicentennial celebrations.[55] In Japan, Sharp took the same approach a little further, releasing one calculator reworked into a folding makeup compact and another incorporated into the cover of a notebook.[56]

Another popular route was to follow Time Computer's lead by bundling calculators together with other devices to fill ever-narrower niches. The humble clock-radio, for instance, was an obvious target for crossbreeding: the fashion designer Pierre Cardin gave his blessing to a pocketable calculator-clock-radio sometime during the 1980s, but such was the deluge of branded items to bear his name that the precise date is lost.[57] A few years earlier, the Hanimex company of Australia released a combined calculator and Dictaphone, unwittingly uniting the pocket calculator with the *other* vehicle that Texas Instruments had once considered for their microchips.[58] And in the late 1970s, Sharp took the concept to a thoroughly meta plane, creating in the ELSI Mate EL-8048 the glorious redundancy of a calculator combined with an abacus.[59]

A still other way to attract new customers was to turn the cal-

---

[*] Triumph-Adler was not the only manufacturer guilty of such blunt marketing. Casio, Texas Instruments, and others also released models targeted at women.

culator's mathematical smarts to new ends. The simplest calculators could have their circuits redesigned to tackle new problems; those driven by programmable ROM chips could be invested with new algorithms; and a new breed of user-programmable pocket calculators was even easier to customize. In the first category were modest offerings such as Mostek's 1975 check calculators, which let users track their spending, and the TI-1260 shoppers' special, which could calculate prices for goods sold by weight.[60] In the second group lay specialist models such as Hewlett-Packard's HP-12C financial calculator, released in 1981 and still available today* in a slightly modified form.[61] And the final group, where anyone could program a calculator and call it a new product, saw some truly original entries. One of the most striking, perhaps, was the U.S. Marine Corps' VSTOL/REST calculator of 1978. A collaboration between McDonnell–Douglas Aircraft and Texas Instruments, this otherwise standard TI-58 contained a removable memory chip that encoded flight operations for the vertical takeoff Harrier aircraft.[62] It is sobering to think that the safe operation of a $24 million fighter jet was largely dependent on the $130 calculator strapped to its pilot's thigh.[63]

Yet all these efforts were as nothing when measured against the single-minded determination of Casio, a rising star of Japan's electronics industry, to sell calculators any and every way it could think of. Having invented the first relay-driven calculator in the form of the 14-A, Casio nevertheless had been slow to understand the promise of the relay's successors, the vacuum tube and the transistor. Desperate, by the mid-1970s, to catch up with its competitors, Casio embarked on a truly remarkable spree of pocket calculators.

---

* The HP-12C is legendary for its robustness as much as its longevity. A delicious but apocryphal story goes that a zookeeper who used an HP-12C to compute feed mixtures accidentally dropped the calculator into a hippo's food. The HP-12C made the journey through the hippo's digestive tract in one piece and, once washed, worked as well as ever.

It began in 1976 with the CQ-1, or "Calculator Quartz." This was a calculator plus alarm clock wrapped in a bar-shaped, pocket-friendly form factor.[64] But it was more than just a calculator and a clock sharing the same display. Though it lacked the HP-01's clever integration between timing and calculation functions, the CQ-1 could nevertheless reckon with values given in hours, minutes, and seconds as well as decimal numbers. It may have looked like an expedient money grab, but the CQ-1 boasted just enough additional functionality to elevate it above similar devices.[65]

If the CQ-1 had emphasized calculation over timekeeping, 1979's VL-1 went the other way, bolting a vestigial calculator onto a miniature synthesizer—the first ever synthesizer, in fact, to be sold on the mass market.[66] The VL-1's calculator mode had very little business being there at all; there was no clever interplay between the synthesizer and the calculator, just a set of pianolike keys that changed from music to math at the flip of a switch.[67] (One wonders whether the members of the Human League, Elastica, or the Fall, all of whom used the VL-1 at one point or another, ever felt compelled to knock out a quick calculation mid-rehearsal.)[68] Casio returned to the calculator-synth arena with 1981's VL-80, which replicated many of the VL-1's features but replaced its piano keyboard with a more familiar calculator layout. The VL-80 was even more rudimentary than the VL-1, but it did have some noteworthy fans: the German band Kraftwerk commissioned a branded version to be given away in conjunction with their single "Pocket Calculator," and issued instructions on how to play some of their most famous tracks on the diminutive Casio.[69]

Casio saw no boundaries as to what could, or should, be combined with a calculator. 1979 brought the fx-190, a scientific calculator with a ruler along one edge and a long, narrow liquid crystal display of questionable utility that could show computed distances relative to that ruler.[70] The QD-100 "Quick Dialer" chirped out telephone dialing tones; the SG-12 "Soccer Game" doubled as a por-

O   The Casio Mini of 1972: a calculator at the sensible end of the Casio spectrum, and one of the first affordable pocket calculators.[71] *CC BY-SA 2.0 image courtesy of Vicente Zorrilla Palau.*

table videogame.[72] But perhaps Casio's crowning achievement was one that harked back to the Kashio brothers' first ever product, the *yubiwa* cigarette pipe. The magnificent QL-10 of 1979 was nothing less than a calculator united with a cigarette lighter.[73] Through the calculator's heyday of the 1970s and 1980s, Casio would design and sell almost two thousand distinct models of calculator.[74]

WHAT OF THE Pulsar Time Computer Calculator? With its tiny buttons and giant price tag, the Pulsar had always been a status symbol rather than a serious tool, and as such, it was not a direct rival to conventional pocket calculators. But Time Computer, Inc., had its own problems, and the LED watch sector it had created had even bigger ones.

Economies of scale, better manufacturing processes, and other technological advancements propelled calculator and LED watch

O   Look on my works, ye mighty: 1979's Casio QL-10 synthesized the company's past and future in idiosyncratic fashion. *CC BY-SA 2.0 image courtesy of Vicente Zorrilla Palau.*

prices on a relentless downward trajectory. By 1977, Texas Instruments was pricing its LED watches at as little as $20; in stores, they could be had for $15.[75] Smaller manufacturers such as Time Computer could not keep up, and that same year, Hamilton got out of the LED watch business entirely. Its Time Computer subsidiary went to an American jewelry firm and the Pulsar name was bought by Hamilton's Japanese nemesis, Seiko.[76]

Ironically, even the semiconductor giants who had muscled their way in to digital watches largely failed to divine where the wind would blow next. In 1973, Seiko released its own take on the digital watch, but the 06LC did not glow with the characteristic red light of the LED—rather, it sported a pale-green display on which the hours, minutes, and seconds were counted out in flat black numerals. This was an LCD, or liquid crystal display, in which molecules realigned themselves in response to electrical fields to reflect or transmit light as required. LCDs were not entirely novel—there had been a handful of LCD watches made in America and Europe during the early 1970s—but Seiko's was the first to make a lasting impression.[77] And because LCDs reflected light rather than emitting it, LCD watches could keep their displays on indefinitely without draining their batteries. By 1980, of the semiconductor companies that had rushed to sell LED watches, only Texas Instruments remained. And by 1981, TI also had admitted defeat, a victim of its all-in bet on LEDs rather than LCDs.[78] It was a new world with no place for dinosaurs like the Pulsar Time Computer Calculator.

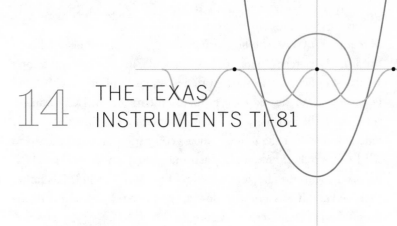

14 THE TEXAS INSTRUMENTS TI-81

B Y 1980, the frenetic pocket calculator race of the preceding decade had slowed to a gentler simmer. America's once-voracious appetite for calculators now fluctuated only by a few percentage points each year, leading the U.S. Department of Commerce to drop the devices from their yearly forecasts of industrial output.[1] The calculator's place in popular culture was under pressure, too, as personal computers made by companies such as Amstrad, Apple, Atari, Commodore, Radio Shack, and Sinclair began to appear in shop windows and on Christmas lists. Kraftwerk might still be singing about their pocket calculators, but in the movies, computers would take starring roles in blockbusters such as *TRON*, *War Games*, and *Superman III*.[2]

In these films and in many other contemporary works of fiction, computers were presented as tools to be used for evil or simply as evil in and of themselves. But skepticism toward technology, and especially information technology, was hardly a new phenomenon. On receiving the gift of writing from the god Thoth, for example, an ancient Egyptian king named Thamus fretted that his subjects' memories would atrophy as they came to rely upon the demon papyrus.[3] For more recent examples, look no further than the *New York Times*'s nineteenth-century excoriation of Alexander Graham Bell's "atrocious" telephone, and the *Spectator*'s lamentation of the "constant diffusion of statements in snippets [and] the constant excitements of feelings unjustified by fact" perpetrated by the telegraph.[4] (Replace "telegraph" with "social media," and it becomes apparent that the same anxieties exist even today.)

Mathematical technologies attracted the same opprobrium. In sixth-century BCE China, Lǎozi, author of one of Daoism's foundational texts, opined that "good mathematicians do not use counting rods."[5] Then came the early modern beef between the abacists with their abacuses and the new-school algorists who used pen and ink to reckon with Hindu-Arabic numerals.[6] Even as late as the nineteenth century, Charles Xavier Thomas's arithmometer could still raise eyebrows at the likes of Spain's Royal Academy of Sciences, whose members worried that users might lose their facility for paper-and-pen math.[7]

For calculator manufacturers in the late 1970s, who saw their sales weakening and their profit margins eroding, there was one frontier left to be exploited. But it was a market for which they would have to fight hard, and fight dirty, because in trying to crack it they would come up against the same age-old prejudices that had stymied many others before them. Pocket calculators were everywhere except the classroom.

TEACHERS AND EDUCATORS, especially in the United States, had started experimenting with calculators almost as soon as prices had crept within the bounds of affordability. By 1976, sufficient experience had been gained, and enough opinions formed, for a teacher named Marilyn Suydam to compile a 377-page report on the use of calculators in schools.[8] Written for the National Science Foundation of Washington, DC, Suydam's report captured the hopes and fears that the electronic calculator elicited in the minds of parents, teachers, and school boards. Chief among them was a concern that introducing calculators to the classroom might erode basic arithmetical skills. As one of Suydam's respondents put it, "There is a real danger that if calculators are used, children will think that pushing buttons on a black box is mathematics."[9] Educational publishers, too, were reluctant to shake things up—although not, it must be said, because of any noble desire to protect the sanctity of America's math education, but rather because updating math textbooks was an arduous and expensive process.[10]

Yet many teachers saw obvious benefits. Less confident students could save time that would otherwise be wasted on simple, repetitive operations, while high achievers could experiment with sophisticated concepts such as logarithms, exponentials, and trigonometric functions. Looking beyond the classroom, it was clear that many adults were already making use of calculators in their daily lives—and it was better, some said, to prepare children for today's world rather than yesterday's.[11]

In a more abstract way, the calculator also offered a cathartic break with the past. The USSR's 1959 launch of *Sputnik*, the first artificial satellite, had sent the USA into a crisis of self-confidence that, in turn, drove a generation of math teachers to embrace the sort of challenging, theoretical mathematics on which the careers of rocket scientists are founded. They called it the "New Math." It was not, unfortunately, the sort of math that school students were very much interested in learning, and standardized test scores suffered accordingly.[12] By the early 1970s, a disgruntled teaching establishment had rebelled so thoroughly against the New Math that the pendulum had swung firmly in the other direction; now, basic, practical arithmetic was the order of the day. But the pocket calculator threatened this new approach: if a simple four-function calculator could lay waste to whole swathes of the K–12 math program, might the curriculum now be *too* simplistic?[13] The arrival of the calculator raised some uncomfortable questions.

As Marilyn Suydam and others wrestled with these issues, time was running out to formulate answers. Suydam cited a 1975 survey that had found that 1 in 5 Ohio schools were already using calculators in the classroom, and that 1 in 3 had forbidden their use in one context or another.[14] The calculator was not about to go away.

I N 1974, Bert Waits and Franklin Demana were math instructors, both in their mid-thirties, at Ohio State University.[15] Their department's intake that year included 4,500 remedial math students with issues, as the faculty saw it, ranging from poor arithmetic to poor attitudes. Together with a band of fellow teachers, Waits and Demana put together a syllabus that introduced calculators as computational aids for their students—and, implicitly, as shiny baubles to hold their attention. To support the program, the OSU math department bought thousands of custom-made, four-function pocket calculators from Texas Instruments and put them on sale in campus bookstores for $16.30 each, or about the same as a textbook.[16]

The initial results were promising. Freed from the need to calculate every last sum and multiplication on paper, OSU's remedial students were more inclined to probe at the theory behind the problems they were set. It was easier to grasp the meaning of a mathematical function, such as $y = mc + c$, if one could simply throw numbers at it and plot the results on graph paper.* Attendance was good, attitudes were good, and performances were good.[17] Waits and Demana became calculator converts.

For its part, Texas Instruments had come a long way from its struggles to make the handheld Cal Tech ready for public consumption. To its line of calculator microchips, TI had added an increasing number of other components until, in 1972, the firm announced its first complete calculator. That device, the handheld TI-2500 Datamath, was followed two years later by the company's first scientific calculator, the SR-50, which aped the groundbreaking HP-35 down to the colors of its keys and its tumblehome casing.[18] TI would ultimately launch more than a hundred different models before the decade was out.[19]

Among this avalanche of calculators were more than a few aimed at the untapped demand in school classrooms. Mostly famously, the mustachioed "Little Professor" of 1976 used calculator technology to teach elementary schoolers about math, with more advanced "Math Magic" and "Wiz-A-Tron" models following the next year.[20] But TI's strategy went far beyond the Little Professor, as eye-catching as it might have been, and by 1977, the company had assembled a lineup of calculators for every stage of a child's education. For the youngest students there was the ABLE calculator, whose interchangeable faceplates started with keys only for ⓪, ①, and ⊞, and which led its user all the way up to ⓪ through ⑨, ⊞, ⊟, and ⊠.[21] Next came the four-function TI-1205 with its simple, bright design; the TI-1255 with an added memory function; the solar-powered TI-1766 to save students from exam-day battery anxiety; the TI-30, an entry-level scientific calculator; and, finally, the SR-50 to accompany stu-

---

* Results that show, of course, a straight line with gradient m passing through the y axis at point c.

dents into college and beyond.[22] This was not merely marketing by carpet-bombing, either. TI had developed the ABLE calculator in conjunction with the University of California, and the same institution produced a teaching program for use with the TI-1205. The University of Denver provided similar help for other models.[23]

To be sure, TI was not the only electronics company chasing the school dollar. In 1974, for instance, Hewlett-Packard donated a clutch of HP-35s to the USA Mathematical Olympiad, a high school mathematics contest.[24] And when Marilyn Suydam surveyed calculator manufacturers for her report, a majority of the respondents said that they advertised to educational buyers, employed staff members to engage with math teachers, or both.[25] But the fact remained that despite the industry's best efforts, at the end of the 1970s fewer than 1 in 30 teachers had calculators available to use in their classrooms.[26]

Gradually, however, the dominoes fell. In 1980, the National Council of Teachers of Mathematics, or NCTM, published guidelines that said that "mathematics programs [should] take full advantage of the power of calculators and computers at all grade levels." That was the third of the NCTM's list of recommendations, but numbers one and two could be read as saying much the same thing: "Problem solving [should] be the focus of school mathematics," and "mathematics [should] be defined to encompass more than computational facility" were coded encouragements to stop worrying so much about pen-and-paper arithmetic.[27] The New Math was back, after a fashion at least, but it would need calculators to make it work.

Across the USA, cities and states began to recommend, requisition, and require calculators in schools. In 1986, for example, Connecticut bought 35,000 TI calculators so that students who could not afford a calculator could still use one.[28] Chicago and New Jersey followed suit two years later, and, in 1992, New York and New Jersey made calculators mandatory for certain math exams.[29] Yet the teachers who had to navigate these sweeping changes were often left without guidance on how to do so, and many parents were still concerned that calculators would stunt their children's ability to do basic math.[30]

B ERT WAITS AND Franklin Demana were not to be cowed. In the early 1980s, the pair extended their calculator experiment into high schools in the form of the "Calculator and Computer Pre-calculus Project." C²PC was a multifaceted effort to prepare high schoolers for math that they were likely to encounter in college, comprising course materials, a textbook, teacher training courses, and even, in the early years, computer software.[31]

Calculus, as you may remember (or, conversely, as you may have chosen to forget), is an occasionally fiendish branch of mathematics pioneered by Gottfried Leibniz and Isaac Newton that studies how expressions such as x^2 or $\sin(x)$ change as one varies their inputs.[32] Mathematically and pedagogically, calculus was a much more challenging subject than the remedial math that had led Waits and Demana to first deploy calculators in the classroom, so this time around the pair employed an even more sophisticated tool—the desktop computer. With the right software, students could plot graphs directly rather than having to plug numbers into their calculators over and over again.[33]

Then, in 1985, a new pocket calculator appeared. It supported the same functions as most other scientific calculators, such as sines, exponentials, logs, and so on, but it could also plot those functions on a monochrome, liquid crystal screen just three inches square.[34] Even better, it cost $69.95, or £79.95 in the U.K.—much, much less than the $8,500 price tag of the earliest-claimed graphing calculator of a decade earlier, the desktop-engulfing Tektronix 31/10.[35]

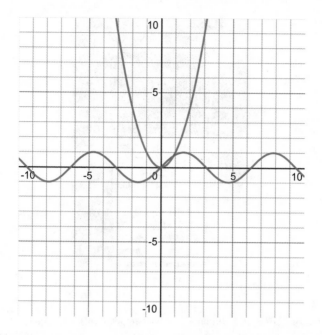

O In red, $y = x^2$. In blue, $y = \sin(x)$. These graphs were generated using Desmos, a free online alternative to graphing calculators. *Image generated with Desmos (https://www.desmos.com/calculator).*

Waits and Demana were agog. Think of the educational possibilities! Almost as soon as the new calculator appeared, the pair abandoned their computer programs in favor of it. In a later report on their work, they acknowledged that "the invention of graphing calculators [. . .] is what made C²PC a viable and implementable course."[36] They were gratified, too, that many of the students taught according to their calculator-based syllabus left with a better understanding of how functions and graphs were related.[37] In 1989, when Bert Waits coauthored a study on the future of math teaching, it was not a surprise to see it recommend that all students in grades nine through twelve should have access to graphing calculators.[38]

Texas Instruments, once Waits and Demana's first port of call

O The Casio fx-7000GA. Its immediate and near-identical predecessor, the
fx-7000G, led Texas Instruments to design and build a rival graphing calculator.
*"Casio Fx-7000GA Handheld Electronic Calculator." National Museum of American
History. Washington, DC: Smithsonian Institution, n.d. https://americanhistory.si.edu/
collections/search/object/nmah_599946. Image courtesy of Division of Medicine and
Science, National Museum of American History, Smithsonian Institution.*

for calculators, was aghast at all of this, because the new device was
not a TI calculator. It was, instead, the Casio fx-7000G, a product of
one of TI's fiercest rivals in the calculator business, and it was a body
blow to TI's plans for the American classroom. Clearly, Something
Had to Be Done.

THE FIRST POINT of business was to build a competing
product. Hewlett-Packard had done just that in 1987 with the

HP-28C, a graphing calculator that boasted an alphanumeric keyboard and a built-in programming language called RPL. But at $235, the 28C was too expensive for most classrooms, and perversely, it was too capable for many of them as well.[39] A contemporary review of the 28C imagined it passing a college-level math exam all by itself, since among its party tricks was the ability to solve many of the same equations that calculus students were expected to tackle with a pencil and paper.[40] The 28C occasioned the same soul-searching in colleges that had occurred in high schools a decade earlier.

Texas Instruments took a different approach. Or rather, it took precisely the same approach that Casio had done in 1985, down to the smallest detail. When the TI-81 debuted in 1990, customers would have been forgiven for wondering if the preceding five years had elapsed at all. The TI model's overall layout, including its 96-by-64-pixel screen, was clearly inspired by that of the fx-7000G, and the American model handled keyboard input very similarly to its Japanese predecessor.[41] The similarities extended inside the TI-81's blue-gray casing. Both its CPU, and that of the Casio, were based on a chip called the Zilog Z80, and both calculators sported near-identical sets of supporting components.[42] A generous observer might have called the TI-81 an homage; a skeptical one, a clone.

A decade old by the time TI incorporated it into its first graphing calculator, the Z80 CPU was the brainchild of Federico Faggin and Masatoshi Shima, who had founded their own chipmaking business after completing the 4004 project at Intel. The Z80 was a simple CPU that owed much to that same Intel chip, and it went on to power many home and office computers, including Radio Shack's TRS-80 and the MSX-branded models that were once common in Europe and Japan.[43] It also had a serious gaming pedigree: not only were franchises such as Metal Gear and Bomberman launched on the MSX, but the Z80 could also be found in arcade game cabinets

O The calculator that launched a thousand mathletes. The Texas Instrument TI-81 was not the first handheld graphing calculator, but it and its successors have become fixtures of the U.S. school system. *"Texas Instruments TI-81 Handheld Electronic Calculator." National Museum of American History. Washington, DC: Smithsonian Institution, 1991. https://americanhistory.si.edu/ collections/search/object/nmah_472820. Image courtesy of Division of Medicine and Science, National Museum of American History, Smithsonian Institution.*

such as Pac-Man and Galaga and games consoles such as Sega's Master System and Coleco's ColecoVision.[44] Cheap and adaptable, the Z80 was the perfect CPU for a graphing calculator aimed at high school students.

O——O

B UT TI WAS NOT CONTENT to let the TI-81 stand on its own merits. For the second prong of its strategy to combat the fx-7000G, the company recruited a pair of familiar voices to sing the praises of their new product. Franklin Demana and Bert K. Waits had already collaborated with TI during the development of the TI-81, and upon its launch they took on new roles as front men for the calculator.[45] They began giving seminars on how the TI-81 could be used to teach C^2PC courses, and, in a 1994 retrospective of the first decade of their project, they noted that they now used TI calculators "almost exclusively."[46]

This public shift in allegiance coincided with a raft of changes for C^2PC. Waits and Demana's training courses were now delivered at the University of Texas at Arlington, and, tellingly, they were now sponsored by Texas Instruments.[47] By 1996, TI had acquired C^2PC lock, stock, and barrel, and the renamed "Teachers Teaching with Technology" (T^3) organization had been moved to Dallas, Texas, so that TI might more readily supervise it.[48] Even beyond T^3, the company was relentless in its courtship of the school dollar: it paid for studies of graphing calculators in secondary schools, funded teacher-training workshops, and reported on how its teaching materials were deployed in practice.[49] At the time of writing, TI has paid for or otherwise supported more than 120 publications on the use of calculators in education in the United States and abroad.[50] All this worked hand in glove with a series of deals with educational publishers that ensured that textbooks gave instructions on how to use TI's models, not Casio's, to solve math problems.[51]

Nor were TI's tactics entirely carrot-based. A lobbying budget that hovered around $2 million per year for much of the '90s and '00s oiled a revolving door between public office and private employment, so that TI's payroll included ex-employees of state and federal governments, political advisors, and even a former state representative.[52] These lobbyists fought against the introduction to schools of calculators with touchscreens, alphabetic keyboards, and internet connectivity—all of which were antithetical to the profitable simplicity of the TI-81 and its successors. In Texas, the company even pushed for all high school students to be required to take Algebra II, a course that required a graphing calculator.[53]

The result was the installation of Texas Instruments at the heart of the American math teaching profession. T[3] instructs some sixty thousand U.S. educators each year on how to teach with TI products, not to mention its branches in Australia and Europe.[54] Lists of permissible devices for college admission tests and other examinations are crowded with TI models.[55] In 2000, TI crowed that it had sold 20 million graphing calculators since the TI-81's debut in 1990; three years later, the figure stood at 25 million.[56] By the mid-2010s, the company was selling upward of 9 out of every 10 graphing calculators in the USA—and, by pushing models still based on the forty-year-old Z80 processor, was making profits of almost $50 on every $100 calculator.[57] It was as if Gordon Moore's eponymous law had flatlined in the mid-'90s. This as math teachers in deprived neighborhoods had to shop for secondhand calculators in pawn shops in order to distribute them to their classes, then buy them back again when they inevitably went missing.[58]

YET DESPITE TI'S BEST EFFORTS, educators were growing restive. It started at the top when, in 2000, the NCTM toned down its bullish guidelines on calculators. Rather than demanding that "calculators should be available to all students at all times," the organization now espoused a "principled approach" to be guided by teachers rather than technology.[59] In Britain, too, a reckoning loomed in 2011 when the new government, led by a Conservative Party whose views toward gadgets such as pocket calculators were exactly as progressive as one might have expected, ordered a review of calculators in schools.[60] More damning still, in 2014 the state of Texas removed mandatory Algebra II lessons from its high school syllabus in spite of fierce lobbying by the state's most prominent high-tech employer.[61]

Nowadays, the threats to TI's hegemony are legion. There are as many Casio calculators as TI models on the list for the USA's standard college admission test.[62] (In 2020, a high schooler penned an op-ed for the *Los Angeles Times* that questioned the propriety of requiring a calculator at all for a test already perceived to be slanted against poorer students.[63]) And despite worries over cheating, computer and smartphone apps that mimic or improve on TI's offerings are starting to replace physical calculators in some classes, schools, and exams.[64] The company's financial reports make it impossible to know for sure, but the opacity of the "Other" category, under which TI reports calculator sales, cannot hide the fact that related profits have dropped by more than $1 billion since 2010. (The comparison is especially unflattering when compared to a companywide *increase* in profits from $3 billion to $5.9 billion over the same period.)[65]

NONE OF THIS is to suggest that calculators in the classroom were or are, necessarily, a bad thing. In 2003, surveying three decades' worth of research, Aimee J. Ellington found that, for mixed- and higher-ability classes, students who were allowed to use calculators during lessons but were forbidden from using them in exams did better than those taught without calculators. That is, their ability to do pencil-and-paper math was *improved* by having previously had access to a calculator. Ellington also found that teachers who wove calculators into their lessons as teaching tools, and not merely as aids to computation, got even better results. And perhaps most important of all, Ellington discovered that students who used calculators simply enjoyed math class that bit more.[66] (The U.K.'s review of calculator use, completed in 2018, found much the same.)[67] In summary, calculators undoubtedly have an important part to play in math education, even if that means suffering the occasional classroom humorist to type out "58008" and turn their calculator upside down.

Nor is the intention to single out Texas Instruments as opposed to any other maker of calculators—or, indeed, any other corporation that has slipped the surly bonds of private ownership and danced on the stock exchange. As a public company with a legal duty to its shareholders, we should not be too surprised if TI has, at times, preferred to pursue profits via lobbying rather than invention. It would not be the first business to have done so.

The rush of innovation in the 1970s that birthed the pocket calculator gave way first to diversification in the 1980s and then, by the 1990s, utter ubiquity: in the Western world, anyone who wanted or needed a calculator almost certainly possessed one. Or, more likely, many of them; rare indeed was the kitchen drawer in the 1990s that did not hide at least one broken calculator, its LCD display clouded or its battery compartment crusted with the life blood of exploded AAAs. The TI-81 was not the last pocket calculator, and nor should it be blamed for the calculator's eventual demise, but its launch signaled the moment at which the calculator lost its luster. Calculators had become a commodity, and one of dwindling importance at that, subject to the same chicanery, profiteering, and general economic shenanigans as any other such thing.

N 1978, seven long years before Casio's futuristic fx-7000G would hit the classroom, a casual observer could have regarded the calculator market and concluded that the future of the world's favorite pocketable electronic device looked rosy. That year saw the unveiling of the TEAL Photon,* the first solar-powered pocket calculator.[1] Casio's LC-78, launched the same year, was one of the first credit card–sized calculators and ran for months on a complement of button cells.[2] And at the opposite end of the spectrum came Hewlett Packard's 41C, a $295 programmable evolution of the HP-35 that came close to being a fully fledged computer—and which, at $1,200 in today's money, was not far off the cost of one either.[3]

But for Harvard sophomore Dan Bricklin, none of this was good enough. Having once worked at Digital Equipment Corporation, one of the largest American computer companies, and now studying for an MBA, Bricklin was frustrated with having to hunt

---

* Its manufacturer, Tokyo Electronic Application Laboratory, promptly went bust, but then what were a few bankruptcies here and there if not a sign of a thriving, competitive market?

down and fix numerical errors in his coursework.[4] He tried instead to crunch his numbers on Harvard's PDP-10 mainframe,* a room-filling behemoth concealed in some unseen lair and accessed, for a fee, via remote terminal, but was chagrined to note that a fellow student named Alan Backus often finished his calculations first.[5] Backus's tool of choice was a programmable calculator, and he was not alone in using one: calculators such as the HP-41C, which excelled at long strings of repetitive calculations, were fast becoming necessities for professionals ranging from Army officers to pilots, oil and gas engineers, epidemiologists, meteorologists, and even forest wardens.[6]

Bricklin had no fear of calculators (he had once dressed up as one at Halloween), yet neither his trusty TI Business Analyst nor Alan Backus's programmable model offered exactly the combination of speed, power, and flexibility that he wanted.[7] He had lived through the transition from slide rules to calculators, and could feel another, similar revolution in the air, but where, he wondered, was the "word processor" for numbers?[8] At DEC, Bricklin had helped develop sophisticated electronic typewriters of the kind that were then taking root in law offices and typing pools, yet the pocket calculator, by comparison, remained steadfastly undisrupted.[9] He daydreamed about building a kind of supercalculator that could recompute values on the fly, with a fighter jet–style heads-up display and a computer mouse (a device barely less exotic than his envisioned whiz-bang display) with which to navigate across a sea of numbers.[10] How hard could it be?

---

* It was on the same PDP-10 that, a few years earlier, a young Bill Gates had created the first ever Microsoft software program. Harvard's PDP-10 was funded by the U.S. Department of Defense, which was unamused that an upstart student had coopted it to bootstrap his fledgling company. Gates was admonished by the university authorities and eventually left Harvard two months shy of receiving his degree.

O———O

REALITY FORCED BRICKLIN to temper his ambition. He was not a mechanical engineer or an electronics wizard, but he did know his way around a computer and began to rough out the essential features of his sci-fi calculator on the PDP-10. His vision gelled over the spring of 1978. He wanted users of his program to be able to see all the figures they were working with at once (quite unlike a calculator, with its single line of digits), and inasmuch as the user might need to enter formulas to calculate sums, averages, compound interest, or whatever else, those formulas had to take a backseat to the numbers themselves.[11] He wanted to make, as he later described it, "a magic piece of paper," but even a piece of paper can benefit from a little structure. To give his program that structure, he borrowed the rows and columns of an accounting tool called a spreadsheet.[12]

Mathematical grids, or tables, were thousands of years old by the time Bricklin arrived on the scene. A historian named Martin Campbell-Kelly has observed that as soon as one has a flat surface, a writing implement, and some sort of repeated calculation to do, a mathematical table is a practical inevitability.[13] The ancient Sumerians' clay tablets of square and cube roots are perfect examples, but Bricklin had in mind a more recent species. Named for the "spreads" of paper on which they were printed, spreadsheets were tables on which credits and debits were totted up to balance a company's books. Eventually, the use of gridded sheets to map out financial computations became so common that preprinted spreadsheets were available off the shelf.[14] But there were limits to the technique. Bob Frankston, Bricklin's eventual partner and collaborator, spoke of grids that sprawled across blackboards and even across multiple rooms, whose creators were doomed to erase and recalculate strings of related figures whenever a single input changed.[15] It would have been a laborious process regardless of whether a calculator was on hand.

That would change in the early 1960s. The computer, with its ability to blaze through mathematical operations, was the ideal tool for mechanizing tiresome, repetitive calculations, but something was lost in the first attempts to do so. The graphical user interfaces that dominate today's computers and smartphones were exceedingly rare at the time, and the spreadsheet's familiar grid was thrown out in favor of esoteric textual commands* and programming languages.[16]

---

* As a taster, what follows is a contemporary "demo" program written in BBL, a business-oriented programming language, that produced a simple income statement. Programs like this represented one possible path through the Choose Your Own Adventure book of business computing; the computerized spreadsheet was another.

```
1000 FIX COLUMNS=7 CHARACTERS=9 COMMA=ON
      DECIMAL=0 LINEREAD=ON
2000 DEFINE REP1(13,5)
2010 SHOW /// ><"BBL DEMO"
2020 SHOW ><"INCOME STATEMENT"
2030 SHOW >>"1ST",>>"2ND",>>"3RD",>>"4TH",>>"TOTAL"
2040 SHOW 4:>>"QT",>>"YEAR"
2050 SHOW /"WIDGETS",ROW1
2060 SHOW "BLEVETS",ROW2
2070 SHOW 5:>>"------"
2080 SHOW " TOTAL SALES",ROW3
2090 SHOW /"COST OF SALES",ROW5
2100 SHOW "G & A",ROW6
2110 SHOW "INTEREST",ROW7
2120 SHOW "DEPRECIATION",ROW8
2130 SHOW "5:>>"------"
2140 SHOW " TOTAL EXP",ROW9
2150 SHOW /"NET IC B4 TAX",ROW10
2160 SHOW "TAXES",ROW12
2170 SHOW "5:>>"------"
2180 SHOW "NET INCOME",ROW13
2190 SHOW "5:>>"------"
2200 REPEND
3000 REMARK - READ FORECAST FOR 4 QUARTERS
3010 READ ROW1
```

The first number-wrangling programs were expensive too: in the 1970s, a financial package designed to run on a mainframe computer could cost as much as a car, and even renting time on such a system might run to $10 or more per hour.[17]

But Bricklin's program was not meant for such rarefied company. Barbara Jackson, a professor to whom he pitched his electronic spreadsheet, told him that he was competing not with mainframe computers but with the back of an envelope: "Senior executives don't

```
3020 READ ROW2
3030 READ ROW4
3040 READ ROW6
3050 READ ROW7
3060 READ ROW8
3070 READ ROW11
3080 ROW3=ROW1+ROW2
3090 ROW5=ROUND(ROW3*ROW4,O)
3100 ROW9=ROW5+ROW6+ROW7+ROW8
3110 ROW10=ROW3-ROW9
3120 ROW12=ROUND(ROW10*ROW11,0)
3130 ROW13=ROW10-ROW12
3200 COL5=COL1+COL2+COL3+COL4
3210 DISPLAY REP1
3300 REMARK WIDGET SALES
3310 DATA 2000,2500,2750,3000
3320 REMARK BLEVET SALES
3330 DATA 3000,3000,3400,3500
3340 REMARK COGS % OF TOTAL SALES
3350 DATA .10,.10,.15,.15
3360 REMARK G&A
3370 DATA 100,100,100,125
3380 REMARK INTEREST
3390 DATA 50, 10, 50, 10
3400 REMARK DEPR.
3410 DATA 25,25,25,25
3420 REMARK TAX RATE
3430 DATA .50,.50,.50,.50
3500 END
```

do this [programming] stuff," she said. "You've got to make it simpler, simpler."[18] His competition, as he saw it, was the $35 TI calculator on his desk.[19] If his project was to stand any chance at all, the hulking PDP-10 on which it ran would have to be replaced by something much cheaper.

O N   A   B I K E   R I D E along the shore of Martha's Vineyard in the summer of 1978, Bricklin decided to turn his prototype into a product.[20] This was when he recruited Bob Frankston, a programmer he knew from business school, and at the suggestion of one his professors, he contacted a Harvard alumnus named Dan Fylstra, who published games and other computer software from an office in his apartment. The three agreed to collaborate on Bricklin's invention: Bricklin and Frankston as a newly incorporated company called Software Arts, and Fylstra via his publishing business, Personal Software.[21]

The first order of business was to decide which machine should run the new program. All three men were steeped in the world of computers, and any lingering thoughts of building a calculator or similar device must have evaporated long before. Dan Fylstra in particular was an enthusiastic booster of what were variously called "home," "personal," or "micro" computers, as distinct from traditional, room-sized mainframes and the fridge-sized "minicomputers" that were gradually replacing them. He showed Bricklin and Frankston what he considered to be the front runners: a Commodore PET, a Radio Shack TRS-80, and an Apple II. The last of these Fylstra had bought directly from Steve Jobs, one of Apple's founding partners, and quite aside from the machine's technical merits, he was convinced that Apple was destined for greater things. Accordingly,

Bricklin borrowed Fylstra's square, beige Apple II and set to work on a new prototype.[22]

It was on that Apple II that Bricklin's electronic spreadsheet came to life—and where his wildest ideas went to die. Out went the heads-up display and in came the Apple's monochrome monitor. Out went the mouse and in came the Apple's "game paddle," a kind

O The Apple II, a vehicle for VisiCalc. *"Apple II Personal Computer." National Museum of American History. Washington, DC: Smithsonian Institution, 1980. https://americanhistory.si.edu/collections/search/object/nmah_334638. Image courtesy of Division of Medicine and Science, National Museum of American History, Smithsonian Institution.*

of primitive joystick for playing games like Pong. The paddle, however, turned out to be too twitchy as a navigation device, so Bricklin fell back to the arrow keys on the Apple's keyboard.[23] But the Apple also enabled new features: its speaker emitted a *thud* whenever the user came to the edge of the spreadsheet, and a *ding* whenever an error occurred.[24]

As Bricklin finessed the program's look and feel, Frankston made sure it would work, and work well, within the Apple's limited memory. Ironically, the little home computer lacked the right tools to create programs for itself, so the pair did all their programming on a mainframe computer. Frankston would write code all night, when the mainframe's time was cheap, and Bricklin would pick up after class, where Frankston had left off.[25]

A S FRANKSTON AND BRICKLIN perfected their program, Dan Fylstra set about selling it. There is an old joke that there are only two hard problems in software engineering: cache invalidation, naming things, and off-by-one errors. At some point along the way, in the early hours at a diner in Cambridge, Massachusetts, Frankston and Fylstra solved the second problem.[26] That was when they settled on the name "VisiCalc" as a contraction of "visible calculator." Which one of them came up with the name is still a matter of dispute, but putting a name to their unfinished program allowed Fylstra to start promoting it in earnest.[27]

Fylstra kicked off the VisiCalc marketing campaign with a cryptic advertisement in the May 1979 issue of *Byte* magazine. At the foot of a spread promoting Personal Software's other products, a single line read: "VISICALC—How did you ever do without it?"[28] Fylstra toured Silicon Valley with VisiCalc diskettes in hand to ask journalists, industry insiders, and potential sales partners the same question.[29] Bricklin and Frankston showed off VisiCalc at conferences such as the 1979 West Coast Computer Faire, held in San Francisco,* and where, two years earlier, Apple had launched the machine on which VisiCalc ran.[30]

It was not always an easy sell. Computers that sat on desks rather than towering over them were viewed as expensive toys for amateur programmers. There was even talk that the market might be played out by 1980, by which time all of those tinkerers willing to buy a home computer would have done so.[31] Accordingly, commercial programs for home computers were rare, and many of them were buggy, of limited usefulness, or hard to use.[32] That may have explained the scene that greeted Frankston in June 1979, when he gave a paper on VisiCalc at the National Computer Conference. Scanning the audience, he counted only two strangers in the room: everyone else was related to, friendly with, or employed by Bricklin, Frankston, or Fylstra. (Both of the interlopers left before Frankston finished his talk.)[33]

If anything, VisiCalc suffered from being almost too revolutionary for its own good. The idea that a single program could calculate a household budget one day, a golf scorecard the next, and a mortgage application the day after flew over many otherwise perceptive heads.† Peter Jennings, one of Fylstra's partners at Personal Software

---

* Rather than drive, the VisiCalc team took the train to the conference. A gasoline shortage precipitated by striking Iranian workers was a reminder that the world still ran on oil, not computers.

† I remember my father introducing me to Lotus 1-2-3, a VisiCalc competitor, in the late 1980s. It ran on a "luggable" Toshiba PC whose screen rendered everything in shades of orange, and as a schoolboy rather than a businessman, I had no idea why this esoteric piece of "magic paper" could possibly be useful. I still used that Toshiba every chance I got, because, even in orange, *Prince of Persia* was and remains a sublime videogame.

(and no relation to the ABC news anchor of the same name), recalled demonstrating a profit-and-loss spreadsheet to bosses at Apple Computer. Mike Markkula, Apple's president, brandished in response a computer tape that bore a checkbook-balancing program—how was VisiCalc any different?[34] Even after VisiCalc had become a resounding success, some users did not realize that it could carry out mathematical operations. Instead, they used it to arrange their numbers in rows and columns and reached for their trusty pocket calculators to recalculate their sums as needed.[35]

The final hurdle to be overcome was a financial one. Despite Frankston's endless tinkering, VisiCalc worked only on the most expensive models of the Apple II, which boasted more than the base version's sixteen kilobytes of memory.[36] Practically speaking, that meant prospective buyers who did not already own a suitable Apple II—that is to say, the vast majority of them—were faced with spending perhaps $5,000 for the privilege of running a $100 software program.[37] Bricklin's aspiration to compete with his $35 calculator was a distant memory.

Yet the team persisted. Frankston continued to refine VisiCalc's code. Bricklin created a reference card that guided users through VisiCalc's features and drafted a more comprehensive manual to go with it. The latter was rewritten first by a freelance writer and then again by Fylstra and Jennings—an unusually thorough effort for a piece of home-computer software. To ship VisiCalc's floppy disk, reference card, and manual, Fylstra and Jennings came up with an embossed brown folder that evoked old money rather than new ideas.[38] And finally, Bricklin learned to tailor his sales pitch to his audience:

> In those days, if you showed it to a programmer, he'd say, "Well, that's neat. Of course computers can do that—so what?" But if you showed it to a person who had to do financial work with real spreadsheets, he'd start shaking and say, "I spent all week doing that." Then he'd shove his charge cards in your face.[39]

```
B5   (U) +B3-B4
Command:  BCDEFGIMPRSTUW-
```

	A	B	C	D	E
1	Year	1979	1980	1981	1982
2					
3	Sales	54321	59753	65728	72301
4	Cost	43457	47802	52583	57841
5	Profit	10864	11951	13146	14460
6					
7					
8					
9					
10					
11					
12					

O The IBM PC version of VisiCalc, released shortly after the original Apple II program. The user interface was largely unchanged. © *Daniel Bricklin, www .bricklin.com.*

GRADUALLY, the message was heard. One of Fylstra's early demos paid dividends when Ben Rosen, an influential computer industry analyst who saw VisiCalc before its public unveiling, wrote that "VisiCalc could some day become the software tail that wags (and sells) the personal computer dog."[40] Rosen was right: in 1982, *InfoWorld* magazine spoke of Apple IIs being sold as "VisiCalc accessories," and the sentiment was echoed by *Byte*, which said that VisiCalc was the first software program to sell entire home computer systems.[41] Not that home computers would be confined to the home for much longer, since, in the wake of VisiCalc's debut, Apple found that most of its computers were now being bought by small businesses.[42] VisiCalc stealthily infiltrated big business too: although home computers were still taboo at a corporate level, many office workers dipped into petty cash to buy an Apple and a copy of VisiCalc for their desks.[43] In the end, perhaps a quarter of all Apple IIs sold in 1979 left stores purely as vehicles to run VisiCalc.[44]

Apple and VisiCalc's combined jaunt into the corporate world did not go unnoticed. IBM, a firm synonymous with capital-M mainframe computers, had once looked down on the scrappy world of the microcomputer. Now it observed with interest, and after a fast-moving, yearlong skunkworks project, it launched the IBM PC in 1981.[45] It was a shot heard around the computer world.

IBM had already tried, and largely failed, to sell a microcomputer. That was 1975's IBM 5100, a forbidding desktop machine that responded only to the same esoteric incantations spoken by the company's gigantic mainframes. Worse, only IBM's own software was approved for use on the 5100. As a machine and as a concept, it was the mirror image of the anarchic free-for-all of the nascent home computer scene.[46] Yet the failure of the 5100 showed IBM what not to do the second time around. With this in mind, and having waited until the dust had settled on the first rush of Apples, Commodores, and Ataris, IBM was able to plot a safe course for the PC through the microcomputer minefield.

First, the PC was a much more welcoming machine than the standoffish 5100. It was made from off-the-shelf components, such as Intel's 8088 processor, a cousin of the seminal 4004.[47] IBM published specifications that would allow anyone to make PC-compatible keyboards, printers, or expansion cards.[48] And, in the spirit of Hewlett-Packard's HP-35 of a decade before, the PC would be the first IBM product to be sold in shops as well as by their business-to-business sales network.

When it came to software, too, the PC was worlds away from its predecessor. Its Microsoft-designed operating system, PC-DOS, was reassuringly similar to an existing home computer OS called CP/M.[49] (In light of later lawsuits, it may have been a little *too* similar.)[50] It allowed users to run whichever programs they wanted and also to create their own using the ubiquitous BASIC programming language. IBM's masterstroke, however, was launching the PC with a full suite of compatible software, including a word processor, an accounting package, an adventure game—and VisiCalc.[51]

IBM had contacted Software Arts late in 1980. After being sworn to secrecy, and in the finest tradition of electronics engineering, Bricklin and Frankston's company took delivery of a wooden board bearing the circuitry of a prototype PC.[52] A year of programming later, during which the breadboard PC had been kept under lock and key, the Software Arts team gathered to hear Bricklin read from IBM's imminent press release. He cautioned his employees not to reveal anything about the PC until IBM had made an official announcement, but partway through the meeting, a voice piped up: "It's on the ticker," someone said, the "ticker" being the live news service subscribed to by Software Arts. The PC, and VisiCalc's pride of place as a founding software application for it, were now public knowledge. "It's on the ticker?" Bricklin asked, to laughter from his staff. "OK. So now you can tell people."[53]

From that point on, VisiCalc and the IBM PC rode a winning wave together. Although IBM would eventually get out of the microcomputer business, the PC's underlying design, or "architecture," has continued to evolve into today's personal computers. Intel and Microsoft, which the PC made into household names, are among the most valuable and influential companies in the world. Software Arts and Personal Software, not so much.

A T FIRST, everything was rosy for VisiCalc's authors and their publisher. Users bought more copies of VisiCalc for the PC than for any other computer, boosting Fylstra's business to stratospheric heights. In 1981, the year the PC launched, Personal Software's income increased fivefold.[54] In 1982, the company raked in

$33.7 million; the next year, more than $43 million. (Sizeable royalties were passed on to Bricklin and Frankston at Software Arts.) In 1984, the *New York Times* wrote that VisiCalc had done "more than any other product to create the personal computer boom." More than 700,000 copies had been sold, making it perhaps the most popular software program that had ever existed.[55]

It all fell apart just as quickly. Even as Personal Software and Software Arts continued to profit mightily from VisiCalc's popularity, fractures were opening between the two companies, culminating in a 1982 lawsuit in which Fylstra sued Bricklin and Frankston for late delivery of a new PC version of VisiCalc.[56] By itself, the spat might have been survivable. But VisiCalc no longer had the personal computer to itself.

Back in the early days of the VisiCalc experiment, a lawyer had advised Bricklin that patenting a software package would be exceedingly difficult, and that only when an invention was embodied in a physical device was the U.S. Patent and Trademark Office likely to give it the nod. VisiCalc, existing as it did in the intangible world of bits and bytes, was almost certainly ineligible.[57] Accordingly, Bricklin had pushed on without a patent—which meant he had no retort to the many "Visiclones" and "Calcalikes" that joined the fray in the years after VisiCalc's debut.[58]

The final insult was the appearance, in 1984, of a PC spreadsheet called Lotus 1-2-3. It had been developed by Mitch Kapor, one of Fylstra's ex-employees, who had listened as Bricklin reeled off what he considered to be VisiCalc's biggest failings.[59] Kapor's program was responsive where VisiCalc stuttered and let users draw graphs and charts at the click of a button. It ate VisiCalc's lunch. Amid the ongoing feud between Bricklin, Frankston, and Fylstra, VisiCalc sales dropped 85 percent in less than a year.[60] Finally, in 1985, Kapor's company bought Software Arts for peanuts and took VisiCalc off the shelves.[61]

The saga of VisiCalc's stunning rise and fractious fall has become practically an oral tradition. Harvard Business School lionizes Dan

Bricklin as "[changing] the way business was done," and has renamed a classroom in his honor.[62] (The room formerly known as Aldritch 108 was where Bricklin first sketched out the VisiCalc user interface.)[63] Hindsight has ennobled VisiCalc as the first "killer app"—a program so useful that users would switch computers for it, or even buy a computer to run it in the first place. Its place in the computing pantheon is assured.[64]

OBSCURED BY ALL THIS DRAMA, the device that Dan Bricklin had taken aim at in 1978 had begun a quiet slide into obsolescense. The pocket calculator that had once grabbed headlines and column inches now shrank from the blinding light of the personal computer.

As the 1970s gave way to the 1980s, it was plain that the calculator was in retreat on multiple fronts. Nowhere was it more obvious than when following the money. In 1980, the U.S. Department of Commerce noted that shipments of "calculating and accounting machines" had declined from a 1974 peak of $1.2 billion to $0.89 billion in 1979.[65] By the end of the 1980s, PC shipments had reached $2 billion and the calculator was all but a footnote.[66]

The calculator's technological relevance ebbed along with its commercial impact. Where once Busicom's collaborations with Mostek and Intel had been feted in the pages of trade magazines, now it was Intel's PC chips that were pored over in minute detail. After all, aside from high-end graphing and scientific models, most calculators were capable only of the same age-old functions—add, subtract, multiply, divide, and perhaps calculate percentages—and there was little need to design new chips to handle them. By way of illustration, in 2011 the journal *Science* published an analysis of the world's computational capacity as expressed in MIPS, or millions of instructions per second, where an "instruction" is the unit of work performed by a general-purpose processing chip.[67] In their paper, Martin Hilbert and Priscila López estimated that in 1986, calculators had embodied more than 41 percent of the world's MIPS, handily outclassing all other computing devices. Home computers, mainframes, videogame consoles, and even supercomputers were just a drop in the calculator's ocean. By 2007, however, PCs and game consoles together accounted for more than 90 percent of the world's computational capacity, with calculators claiming one thirty-third of 1 percent.[68]

Worst of all in the post-VisiCalc world, fewer and fewer people seemed to *care* about the pocket calculator. The calculator's golden age could be seen in the bell curve of books, articles, and papers that took the calculator as their subject, which accelerated upward in the early 1970s, crested at more than 350 publications per year in the following decade, and which dropped just as rapidly to a background hum in the early 1990s.[69] The rise and fall in the popularity of books such as *The Calculator Revolution: How to Make the Most of Your Calculator*, *The Electronic Calculator in Business, Home, and School*, and *The Calculating Book: Fun and Games with Your Pocket Calculator* tracked exactly the mindshare afforded to their subject.[70] As time went by, the calculator increasingly succumbed to the Wildean fate of simply not being talked about.

EPILOGUE

ORIS JOHNSON, who was, at the time of writing, Britain's most controversial prime minister for many years, has been wrong about a lot of things. But in a career built on questionable opinions and outrageous pronouncements, one in particular of Johnson's proclamations hits close to home for fans of the pocket calculator. In a 1988 column for London's *Daily Telegraph*, the prime-minister-to-be bemoaned the acquisition of a collection of pocket calculators by the Whipple Museum of the History of Science, in Cambridge, England. Johnson opined satirically that the museum should "branch out from mere science and become a major tourist attraction for its peerless collection of obsolete gadgets of every kind."[1]

It was a cruel way to describe a device that, two decades earlier, had helped usher in the world of ubiquitous computing that we inhabit today, but even at the time Johnson's words had a ring of truth about them. In comparison to the digital watch, for example, or the home computer, or the cellphone, the pocket calculator's day in the sun was brief. Half a century after the first electronic pocket calculators arrived, those once-ubiquitous devices can seem like a distant memory.

What happened, exactly, to the pocket calculator? Its disappearance as an artifact worthy of note has been so complete that it is difficult to say with any certainty, but the sheer pace of techno-logical change may well have had something to do with it. During the 1970s, electronics, computing, and software were evolving so rapidly that the calculator was caught up in a jumble of cause and effect. For a flavor of that foment, consider that Apple's co-founders, Steve Wozniak and Steve Jobs, met at HP during that company's push into calculators. The Steves' early microcomput-ers, the Apple and Apple II, helped spark the home-computer revolution, which in turn gave Dan Bricklin the opportunity to perfect the first computerized spreadsheet. Mitch Kapor, who had worked for Bricklin's publisher, stole the spreadsheet crown with his own program, Lotus 1-2-3, and with it gave the IBM PC one of its first killer apps. That PC, of course, was powered by an Intel CPU—a chip made by the same company whose deal with Busicom, a Japanese calculator manufacturer, had saved the American firm and paved the way for the founding of Silicon Val-ley. Masatoshi Shima and Federico Faggin, two of the engineers who worked on the Intel–Busicom tie-up, went on to design their own chipset that in turn made its way into countless models of pocket calculator.

If it is hard to untangle who influenced whom, or which prod-ucts foreshadowed which others, it is rather easier to pick out some winners and losers. VisiCalc might have been the first software pro-gram to yank the limelight away from the pocket calculator, but it was far from the only one: VisiCalc, Lotus 1-2-3, Microsoft Excel, and other electronic spreadsheets were bona fide game-changers, blowing away the cobwebs in sales departments and accounting firms that had only recently graduated from comptometers to cal-culators. There were similar revolutions in science and technology, too, where free-form math programs such as Matlab gifted scientists the same design-refine-repeat workflow that made spreadsheets so attractive and pocket calculators seem so clumsy. In offices and laboratories around the world, the calculator moved from the desk to the desk drawer.

But the computer was not finished with the pocket calculator. Moore's Law saw to that. By the early 1990s, a new breed of "palmtop" PCs, barely larger than the first handheld calculators, started to appear. Some came from companies native to the computer world, such as Apple and Atari; others from firms such as HP, Casio, and Sharp, with roots closer to the calculator. Mostly, palmtops majored in data management rather than number wrangling: address books, calendars, and word processors were the features that Filofax-weary businesspeople wanted in their expensive new toys. Even so, many palmtops could run electronic spreadsheets. And *all* of them boasted built-in calculators.[2]

O Atari's Portfolio, introduced in 1989. Like the Busicom 141-PF of almost two decades earlier, the Portfolio was powered by an Intel CPU and backed by programmable ROM memory. *Bauer, Martin. "Handdator." Stockholm: Tekniska Museet, 1989. https://digitaltmuseum.se/021026359221/handdator. CC BY image courtesy of Algin, Ellinor / Tekniska Museet.*

Cellphones opened a third front. They would, in time, converge with palmtops and PCs to give us smartphones such as the IBM Simon, the Nokia Communicator, and the Handspring Treo (all, of course, loaded with calculator functions), but even some decidedly un-smart phones such as 1997's Nokia 6110 came with embedded calculators.[3] After all, touch-tone telephones had been using calculator-inspired keypads for decades, so why not wedge a calculator into a cellphone and return the favor?[4]

Desk by desk and pocket by pocket, then, the calculator was driven out by more modern, more capable devices. Until at some point—call it 1990 or thereabouts—society's collective interest in pocket calculators dropped below a detectable threshold.[5] But the

○ The Nokia Communicator of 1998, emblematic of early smartphones. The clamshell case closed to reveal a numeric keypad and a small display to mimic a more standard cellphone—though ironically, the Communicator's calculator application was accessed via the main screen and QWERTY keyboard.[6] *"Mobiltelefon, Dummy." Stockholm: Tekniska Museet, 1998. https:// digitaltmuseum.se/021026359221/handdator. CC BY image courtesy of Nord, Truls / Tekniska Museet.*

pocket calculator was not, and is not, dead. At least, not in the conventional sense.

First, there are the zombies. Rummage around for long enough in any office, home, shop, or restaurant, and you are more than likely to turn up a calculator or two. Maybe one inhabits a desk drawer or an old filing cabinet, underused but still functional, or perhaps it lives beside a cash register or a landline telephone, to be called upon occasionally to total up a takeaway order or to amend a check. Historian Martin Campbell-Kelly accounts for these survivors by observing that for years, calculators were dramatically overengineered.[7] Unlike smartphones, with their fragile touchscreens, ten-billion-transistor CPUs, and ever-changing roster of applications, calculators are built on decades-old technologies and sport a user interface that has changed very little over that same period. If you bought, inherited, or stole a pocket calculator in the past thirty years, and have managed to avoid running it over with a car or dropping it from a height of greater than three feet, there is very little reason to have sought out a new one since then.

The pocket calculator survives in another way, too, having played a trick altogether craftier than merely hanging on through sheer indestructibility. Iain M. Banks, the late sci-fi author, had the perfect word for it. In his argot, the pocket calculator has *sublimed*: it has not died out so much as ascended to a new form of existence. Today, when we reach for a calculator, we first pick up our smartphone or open our laptop computer and only then do we start our favorite calculator app. The first step towards this exalted plane was taken in 1970, when Bell Labs researchers Bob Morris and Lorinda Cherry wrote a program called "dc" for their new PDP-11 computer.[8] Their "desk calculator" was one of the first programs written for Bell Labs' new "Unix" operating system, and it gave users a simple text-based interface capable of arithmetic, percentages, exponentials, and square roots. At the same time that the pocket calculator was being willed into existence at Texas Instruments and beyond, dc was showing the way to a future in which the pocket calculator was everywhere and nowhere at once.

And so it is today. The pocket calculator has become a ghost in the machine—a software echo of a device that, to borrow a phrase, once wagged the hardware dog. The calculator is dead; long live the calculator.

ACKNOWLEDGMENTS

THERE IS, EVEN NOW, a devoted community of calculator users and collectors, many of whom have been generous in helping me understand the calculator's birth, life, and afterlife. In particular, I would like to thank Karen Greenhaus, Jim Hughes, Piotr Samulik, Nigel Tout, and Joerg Woerner for giving me their time.

The calculator also has many eminent historians. I was fortunate to be able to draw on the expertise of William Aspray, Martin Campbell-Kelly, Paul Ceruzzi, David Alan Grier, and Jörn Lütjens, who patiently answered my questions and reviewed my manuscript. Yet others were there to play their own parts in the calculator's story: Dan Bricklin, Ben Wood, and Frank Canova were all gracious in elucidating one aspect or another of the history of the calculator.

Many people helped me compile the photographs and other images in the book. Marcin Wichary sent me a treasure trove of images culled from his research on keyboards. (His book on the subject, *Shift Happens*, will be published around the same time as this one. It promises to be a terrific read.) Magnus Hagdorn, Vicente Zorrilla Palau, Igor Tesic, and Nathan Zeldes were kind enough to let me use some of their own images, while Casio's Shoko Akita, Jenina Bas at the Schøyen Collection, Giovanni Cella at the Museo Nazionale della Scienza e della Tecnologia, Terre Heydari and Ada Negraru at the DeGolyer Library, David Morris at the McGregor

Museum, Dale Philbrick of Texas Instruments, Intel's Leslie Skinner, Alfred Wegener at the Heinz Nixdorf MuseumsForum, and Eleni Zavvou and Eirene Choremi at the Epigraphic Museum of Athens helped me locate images held by their respective organizations. Kay Peterson at the National Museum of American History deserves special thanks for helping me with a whole clutch of images, as do Sara Frankel and Maureen Ton of Harvard's Collection of Historical Scientific Instruments.

My agent, Laurie Abkemeier; my editor, Brendan Curry; my copyeditor, Rachelle Mandik; and Caroline Adams, Anna Oler, and Susan Sanfrey of W. W. Norton all provided much-needed help and support in writing, editing, and producing the book. Milan Bozic and Judith Abbate designed the cover and the interior of the book, respectively, and Joe Lops typeset it.

My wife, Leigh, helped and encouraged me throughout the writing of this book. Our sons, Ivor and Casper, did not help one bit, but they more than made up for it by being super adorable.

Thank you all!

NOTES

INTRODUCTION

1   K. Spindler, *The Man in the Ice* (Orion, 2013); W. Kutschera et al., "Radiocarbon Dating of Equipment from the Iceman," in *The Iceman and His Natural Environment* (2000): 1–9, https://doi.org/10.1007/978-3-7091-6758 -8_1.

2   Jessanne Collins, "The History of Fanny Packs and Waist Bags," Quartz, December 13, 2018, https://qz.com/quartzy/1493093/fanny-packs-are-one-of -humankinds-oldest-accessories/.

3   Margaret Dykens and Lynett Gillette, "Opossum," San Diego Natural History Museum, accessed January 17, 2021, https://www.sdnhm.org/ exhibitions/fossil-mysteries/fossil-field-guide-a-z/opossum/.

4   Jessica F. Cantlon et al., "The Origins of Counting Algorithms," *Psychological Science* 26, no. 6 (2015): 853–865, https://doi.org/10.1177/0956797615572907.

5   Russell P. Balda, "Corvids in Combat: With a Weapon?," *The Wilson Journal of Ornithology* 119, no. 1 (2007): 100–102; Barbara C. Klump et al., "Context-Dependent 'Safekeeping' of Foraging Tools in New Caledonian Crows," *Proceedings: Biological Sciences* 282, no. 1808 (2015): 1–8; Bernd Heinrich, "An Experimental Investigation of Insight in Common Ravens (Corvus Corax)," *The Auk* 112, no. 4 (1995): 994–1003.

6   Alice M. I. Auersperg and Auguste M. P. von Bayern, "Who's a Clever Bird—Now? A Brief History of Parrot Cognition," *Behaviour* 156, no. 5–8 (n.d.): 391–407, https://doi.org/10.1163/1568539X-00003550; O. Koehler, "„Zähl"-Versuche an Einem Kolkraben Und Vergleichsversuche an Men-

schen," *Zeitschrift Für Tierpsychologie* 5, no. 3 (1943): 575–712, https://doi.org/10.1111/j.1439-0310.1943.tb00665.x.

7   Irene M. Pepperberg and Jesse D. Gordon, "Number Comprehension by a Grey Parrot (Psittacus Erithacus), Including a Zero-like Concept," *Journal of Comparative Psychology* 119, no. 2 (May 2005): 197–209, https://doi.org/10.1037/0735-7036.119.2.197; K. Plofker, *Mathematics in India* (Princeton University Press, 2008), 54–56.

8   Elliot Collins, Joonkoo Park, and Marlene Behrmann, "Numerosity Representation Is Encoded in Human Subcortex," *Proceedings of the National Academy of Sciences* 114, no. 14 (April 4, 2017): E2806–15, https://doi.org/10.1073/PNAS.1613982114.

9   Christian Agrillo et al., "Evidence for Two Numerical Systems That Are Similar in Humans and Guppies," *PloS One* 7, no. 2 (2012): e31923, https://doi.org/10.1371/journal.pone.0031923.

10   Zhanna Reznikova and Boris Ryabko, "Numerical Competence in Animals, with an Insight from Ants," *Behaviour* 148, no. 4 (2011): 405–434, https://doi.org/10.1163/000579511X568562.

11   Fiona R. Cross and Robert R. Jackson, "Representation of Different Exact Numbers of Prey by a Spider-Eating Predator," *Interface Focus* 7, no. 3 (2017): 20160035, https://doi.org/10.1098/rsfs.2016.0035.

12   Brian Butterworth, C. R. Gallistel, and Giorgio Vallortigara, "Introduction: The Origins of Numerical Abilities," *Philosophical Transactions of the Royal Society B: Biological Sciences* 373, no. 1740 (February 19, 2018): 20160507, https://doi.org/10.1098/rstb.2016.0507.

13   Koleen McCrink and Karen Wynn, "Large-Number Addition and Subtraction by 9-Month-Old Infants," *Psychological Science* 15, no. 11 (November 2004): 776–781, https://doi.org/10.1111/j.0956-7976.2004.00755.x.

1. THE HAND

1   Francesco D'Errico et al., "Early Evidence of San Material Culture Represented by Organic Artifacts from Border Cave, South Africa," *Proceedings of the National Academy of Sciences* 109, no. 33 (August 14, 2012): 13214–13219, https://doi.org/10.1073/PNAS.1204213109.

2   Peter B. Beaumont, Hertha de Villiers, and J. C. Vogel, "Modern Man in

Sub-Saharan Africa Prior to 49,000 Years BP: A Review and Evaluation with Particular Reference to Border Cave," *South African Journal of Science* 74, no. 11 (1978): 409–419.

3   Karl Menninger, *Number Words and Number Symbols: A Cultural History of Numbers* (Cambridge: MIT Press, 1969), 223–248.

4   Andrew Robinson, *The Story of Writing : Alphabets, Hieroglyphs & Pictograms* (London: Thames & Hudson, 2001), 71; D'Errico et al., "Early Evidence of San Material Culture."

5   Justus Möser, Jenny von Voigts, and Friedrich Nicolai, *Patriotische Phantasien* (Berlin, 1776).

6   D'Errico et al., "Early Evidence of San Material Culture."

7   Melissa A. Toups et al., "Origin of Clothing Lice Indicates Early Clothing Use by Anatomically Modern Humans in Africa," *Molecular Biology and Evolution* 28, no. 1 (January 2011): 29–32, https://doi.org/10.1093/molbev/msq234.

8   Menninger, *Number Words*, 7.

9   Menninger, *Number Words*, 9–16; Carl Benjamin Boyer, *A History of Mathematics* (New York: John Wiley and Sons, 1968), 3.

10   Georges Ifrah and David Bellos, *The Universal History of Numbers: From Prehistory to the Invention of the Computer* (New York: Wiley, 2000), 7.

11   Menninger, *Number Words*, 30; Ronald Gatty, *Fijian-English Dictionary: With Notes on Fijian Culture and Natural History*, 2009, 68, 122, https://ecommons.cornell.edu/handle/1813/28702.

12   "Japanese Counter Words: A Complete Guide to Counting in Japanese," Nihongo Master, November 15, 2016, https://www.nihongomaster.com/blog/complete-guide-japanese-counter-words/.

13   Howard Eves, *An Introduction to the History of Mathematics* (New York: Rinehart, 1953), 7; Boyer, *History of Mathematics*, 3.

14   Pim Verheyen, "The Importance of the Body in the Primitive Counting of Tribes" (Universiteit Antwerpen, 2012), https://www.academia.edu/4166555/The_Importance_of_the_Body_in_the_Primitive_Counting_of_Tribes.

15   W. C. Eells, "Number Systems of the North American Indians," *The American Mathematical Monthly* 20, no. 9 (November 1913): 263, https://doi.org/10.2307/2974059.

16   "The Cambridge Anthropological Expedition to Torres Straits and Sarawak," *The Geographical Journal* 14, no. 3 (1899): 302–306.

17   Alfred C. Haddon, "The Ethnography of the Western Tribe of Torres Straits," *The Journal of the Anthropological Institute of Great Britain and Ireland* 19 (1890): 305.

18   Kay Owens, "The Work of Glendon Lean on the Counting Systems of Papua New Guinea and Oceania," *Mathematics Education Research Journal* 13, no. 1 (2001): 47–71, https://doi.org/10.1007/BF03217098.

19   Jürg Wassmann and Pierre R. Dasen, "Yupno Number System and Counting," *Journal of Cross-Cultural Psychology* 25, no. 1 (March 27, 1994): 78–94, https://doi.org/10.1177/0022022194251005.

20   Miguel Civil, "Modern Brewers Recreate Ancient Beer," *Oriental Institute News & Notes* 132 (1991): 1–11; Solomon H. Katz and Mary M. Voight, "Bread and Beer: The Early Use of Cereals in the Human Diet," *Expedition* 28, no. 2 (1986): 23–34, https://www.penn.museum/sites/expedition/bread-and-beer/.

21   Dietz O. Edzard, "History of Mesopotamia," *Encyclopaedia Britannica* (Chicago: Encyclopaedia Britannica, 2009), http://www.britannica.com/EBchecked/topic/376828/history-of-Mesopotamia.

22   Denise Schmandt-Besserat, "How Writing Came About," *Zeitschrift Für Papyrologie Und Epigraphik* 47 (1982): 1–5.

23   Denise Schmandt-Besserat, *Before Writing: From Counting to Cuneiform, Volume 1* (University of Texas Press, 1992), 8; Schmandt-Besserat, "How Writing Came About"; Denise Schmandt-Besserat, "Tokens: Their Significance for the Origin of Counting and Writing," March 2, 2014, https://sites.utexas.edu/dsb/tokens/tokens/; Stephen J. Lieberman, "Of Clay Pebbles, Hollow Clay Balls, and Writing: A Sumerian View," *American Journal of Archaeology* 84, no. 3 (1980): 343, https://doi.org/10.2307/504711.

24   Schmandt-Besserat, "How Writing Came About."

25   J. J. O'Connor and E. F. Robertson, "Babylonian Numerals," MacTutor (School of Mathematics and Statistics, University of St. Andrews, 2000), https://mathshistory.st-andrews.ac.uk/HistTopics/Babylonian_numerals/.

26   Pierre Amiet, "Il y a 5000 Ans Les Élamites Inventaient l'Écriture," *Archeologia* 12, no. 20 (1966): 2; Miguel Valério and Silvia Ferrara, "Numeracy at the Dawn of Writing: Mesopotamia and Beyond," *Historia Mathematica*, 2020, https://doi.org/10.1016/j.hm.2020.08.002; Jöran Friberg, "Three Thousand Years of Sexagesimal Numbers in Mesopotamian Mathematical Texts," *Archive for History of Exact Sciences* 73, no. 2 (2019): 183–216, https://doi.org/10.1007/s00407-019-00221-3.

27 S. L. Macey, *The Dynamics of Progress: Time, Method, and Measure* (Athens: University of Georgia Press, 2010), 92.

28 Menninger, *Number Words*, 201.

29 Ifrah and Bellos, *Universal History of Numbers*, 55.

30 John Roberts, ed., "Pliny the Elder," *Oxford Dictionary of the Classical World* (Oxford University Press, 2007), http://www.oxfordreference.com/10.1093/acref/9780192801463.001.0001/acref-9780192801463-e-1749.pl

31 Pliny the Elder, "That There Were Statuaries in Italy Also at an Early Period," in *The Natural History*, trans. John Bostock and B. A. Riley, vol. 34 (London: Taylor and Francis, 1855), http://data.perseus.org/citations/urn:cts:latinLit:phi0978.phi001.perseus-eng1:34.16.

32 James Grout, "The Temple of Janus (Janus Geminus)," *Encyclopaedia Romana*, accessed February 28, 2021, https://penelope.uchicago.edu/~grout/encyclopaedia_romana/imperialfora/nerva/geminus.html; James Grout, "Roman Calendar," Encyclopaedia Romana, accessed February 28, 2021, https://penelope.uchicago.edu/~grout/encyclopaedia_romana/calendar/romancalendar.html.

33 Plutarch, William Watson Goodwin, and Ralph Waldo Emerson, *Plutarch's Morals: Translated from Greek by Several Hands*, vol. 1 (Boston: Little, Brown and Company, 1874), 189.

34 "Myriad, n. and Adj.," OED Online, June 2003, https://www.oed.com/view/Entry/124528.

35 William Smith, William Wayte, and G. E. Marindin, "frumenta´riae leges," *A Dictionary of Greek and Roman Antiquities* (London: John Murray, 1890).

36 Adolf Berger, "Encyclopedic Dictionary of Roman Law," *Transactions of the American Philosophical Society* 43, no. 2 (March 1, 1953): "tessera nummaria," https://doi.org/10.2307/1005773.

37 Ifrah and Bellos, *Universal History of Numbers*, 55.

38 J. Campbell, "Bede [St Bede, Bæda, Known as the Venerable Bede] (673/4–735), Monk, Historian, and Theologian," *Oxford Dictionary of National Biography*, 2008, https://doi.org/10.1093/ref:odnb/1922.b

39 H. Thurston, "Christian Calendar," in *The Catholic Encyclopedia*, vol. 3 (New York: Appleton, 1908), https://www.newadvent.org/cathen/03158a.htm.

40 H. Thurston, "Easter Controversy," in *The Catholic Encyclopedia*, vol. 5 (New York: Appleton, 1909), 326–327, https://www.newadvent.org/cathen/05228a.htm.

41   Bede and Faith Wallis, *Bede: The Reckoning of Time* (Liverpool: Liverpool University Press, 1999), 9.

42   Bede and Wallis, *Bede,* 10.

43   Ifrah and Bellos, *Universal History of Numbers,* 55.

44   Bede and Wallis, *Bede,* 254–263.

45   Menninger, *Number Words,* 217–219.

46   Ifrah and Bellos, *Universal History of Numbers,* 55, 58–59; Leon J. Richardson, "Digital Reckoning Among the Ancients," *The American Mathematical Monthly* 23, no. 1 (January 1916): 8, https://doi.org/10.2307/2972133.

47   Menninger, *Number Words,* 211.

2. THE ABACUS

1   Peter Strom Rudman, *How Mathematics Happened: The First 50,000 Years* (Amherst, NY: Prometheus Books, 2007), 31–32.

2   Ifrah and Bellos, *Universal History of Numbers,* 11–12.

3   Denise Schmandt-Besserat, "The Token System of the Ancient Near East: Its Role in Counting, Writing, the Economy and Cognition," in *The Archaeology of Measurement: Comprehending Heaven, Earth and Time in Ancient Societies,* ed. Iain Morley and Colin Renfrew (Cambridge: Cambridge University Press, 2010), 27–34, https://doi.org/10.1017/CBO9780511760822.006.

4   A. Leo Oppenheim, "On an Operational Device in Mesopotamian Bureaucracy," *Journal of Near Eastern Studies* 18, no. 2 (1959): 121–128; Schmandt-Besserat, "How Writing Came About," 2.

5   Denise Schmandt-Besserat, "The Envelopes That Bear the First Writing," *Technology and Culture* 21, no. 3 (April 6, 1980): 357–385, https://doi.org/10.2307/3103153.

6   Schmandt-Besserat, "Envelopes"; Lieberman, "Clay Pebbles."

7   Lieberman, "Clay Pebbles"; Menninger, *Number Words,* 166–167.

8   Boyer, *History of Mathematics,* 30–32; O'Connor, Robertson, "Babylonian Numerals."

9   "Introduction: What Is a Lexical List?," Digital Corpus of Cuneiform Lexical Texts (University of California, Berkeley), accessed April 9, 2021, http://oracc.museum.upenn.edu/dcclt/intro/lexical_intro.html.

10   Lieberman, "Clay Pebbles"; "The Pennsylvania Sumerian Dictionary"

(University of Pennsylvania Museum of Anthropology and Archaeology), lu [PERSON], accessed April 8, 2021, http://psd.museum.upenn.edu/.

11   Lieberman, "Clay Pebbles," 350.

12   David Eugene Smith, *History of Mathematics*, vol. 1 (New York: Dover, 1958), 40; "Pennsylvania Sumerian Dictionary," šid [COUNT].

13   C. Walker, *Cuneiform*, Reading the Past (Berkeley, CA: University of California Press, 1987), 7–11.

14   Denise Schmandt-Besserat, "An Archaic Recording System in the Uruk-Jemdet Nasr Period," *American Journal of Archaeology* 83, no. 1 (1979): 22, 27, https://doi.org/10.2307/504234.

15   Lieberman, "Clay Pebbles," 355.

16   Ettore Carruccio and Isabel Quigly, *Mathematics and Logic in History and in Contemporary Thought* (Chicago: Aldine, 1964), 14.

17   Ifrah and Bellos, *Universal History of Numbers*, 126–128.

18   Samuel Noah Kramer, *The Sumerians: Their History, Culture, and Character* (Chicago: University of Chicago Press, 1963), 59–61.

19   "What Is a Lexical List?"

20   Ifrah and Bellos, *Universal History of Numbers*, 139–141.

21   Jona Lendering, "Babylonian Empire," Livius, October 12, 2020, https://www.livius.org/articles/place/babylonian-empire/.

22   Jens Høyrup, "A Note on Old Babylonian Computational Techniques," *Historia Mathematica* 29, no. 2 (May 1, 2002): 193–198, https://doi.org/10.1006/hmat.2002.2343.

23   *The Encyclopaedia Britannica*, vol. XVI (Cambridge: Cambridge University Press, 1911), Letronne, Jean Antoine; A. R. Rangabé, "Lettre De M. Rangabé A M. Letronne Sur Une Inscription Grecque Du Parthénon; Sur Les Peintures Du Théséum Et Des Propylées; Et Sur Deux Monuments Inédits Récemment Découverts. (Suite Et Fin)," *Revue Archéologique* 3, no. 1 (March 27, 1846): 293–304.

24   "Alexandros Rizos Rangavis," Aiora Press, accessed March 27, 2021, https://aiorabooks.com/2019/05/24/alexandros-rangavis/; Alexander Kitroeff, "Memoirs of First Greek-US Ambassador Offer Glimpse Into Post-Civil War America," *The Pappas Post* (Gregory C. Pappas, July 29, 2020), https://www.pappaspost.com/memoirs-of-first-greek-us-ambassador-offer-glimpse-into-post-civil-war-america/.

25   Mabel Lang, "Herodotos and the Abacus," *Hesperia: The Journal of the American School of Classical Studies at Athens* 26, no. 3 (April 16, 1957): 271–288, https://doi.org/10.2307/147100.

26   Herodotus, *The Histories*, trans. A. Godley (Cambridge, MA: Harvard University Press, 1920), 2.36.4, http://www.perseus.tufts.edu/hopper/text?doc=urn:cts:greekLit:tlg0016.tlg001.perseus-eng1.

27   Keith F. Sugden, "A History of the Abacus," *The Accounting Historians Journal* 8, no. 2 (April 16, 1981): 3–5; "Papyrus Sallier 4 (EA10184,6)," British Museum, accessed April 16, 2021, https://www.britishmuseum.org/collection/object/Y_EA10184-6.

28   Demosthenes, *Demosthenes with an English Translation*, trans. C. A. Vince and J. H. Vince (Cambridge, MA: Harvard University Press, 1926), 18.227-231, http://data.perseus.org/citations/urn:cts:greekLit:tlg0014.tlg018.perseus-eng1:227; John Frederic Dobson, *The Greek Orators* (London: Methuen, 1919), 203.

29   Marie-Christine Villanueva-Puig, "Le Vase Des Perses. Naples 3253 (Inv. 81947)," *Revue Des Études Anciennes. Tome 91, 1989, N°1-2. L'or Perse et l'histoire Grecque, Sous La Direction de Raymond Descat*, 1989, https://doi.org/10.3406/rea.1989.4387.

30   "IG II² 2777," PHI Greek Inscriptions (Packard Humanities Institute), accessed April 17, 2021, https://epigraphy.packhum.org/text/5029.

31   Henry George Liddell et al., "A Greek–English Lexicon" (Oxford: Clarendon Press, 1940), ὀβολός , ὁ; Marcus Niebuhr Tod, "The Greek Numeral Notation," *The Annual of the British School at Athens* 18 (April 19, 1911): 98–132.

32   Tod, "Greek Numeral Notation."

33   Menninger, *Number Words*, 300.

34   Lang, "Herodotos and the Abacus."

35   Lang, "Herodotos and the Abacus," 275.

36   Reviel Netz, "Counter Culture: Towards a History of Greek Numeracy," *History of Science* 40 (January 2002): 327.

37   James William Gilbart, *The History and Principles of Banking* (London: Longman, Rees, Orme, Brown, Green and Longman, 1837), 3–4.

38   Menninger, *Number Words*, 315–16.

39   Liddell et al., "A Greek–English Lexicon," ἄβαξ [α^], α^κος, ὁ; Menninger, *Number Words*, 301.

40   J. M. Pullan, *The History of the Abacus* (New York: F. A. Praeger, 1969), 17.

41   Liddell et al., "A Greek–English Lexicon," ἄβαξ [α^], α^κος, ὁ; Menninger, *Number Words*, 301; Pullan, *History of the Abacus*, 17.

42   Yuri Pines, "The Warring States Period: Historical Background," in *The Oxford Handbook of Early China*, ed. Elizabeth Childs-Johnson (Oxford: Oxford University Press, 2021), 579–594, https://doi.org/10.1093/oxfordhb/9780199328369.013.26.

43   Lois Mai Chan, "The Burning of the Books in China, 213 B.C.," *The Journal of Library History* 7, no. 2 (1972): 101–108.

44   "Han Dynasty," *Encyclopaedia Britannica* (Chicago: Encyclopaedia Britannica, 2013), http://www.britannica.com/EBchecked/topic/253872/Han-dynasty.

45   Tsuen-Hsuin Tsien, "Paper and Printing," in *Science and Civilisation in China*, ed. Joseph Needham, vol. 5–1 (Cambridge: Cambridge University Press, 1985), 38–42; Dard Hunter, *Papermaking: The History and Technique of an Ancient Craft*, Dover Books Explaining Science (New York: Dover Publications, 1978), 50–53; Joseph Needham and Ling Wang, "Physics," in *Science and Civilisation in China*, vol. 4–1 (Cambridge: Cambridge University Press, 1962), 230.

46   Tsien, "Paper and Printing," 201–205; Joseph Needham and Ling Wang, "The Gunpowder Epic," in *Science and Civilisation in China*, vol. 5–7 (Cambridge: Cambridge University Press, 1986), 1.

47   Joseph Needham and Ling Wang, "Mathematics and the Sciences of the Heavens and the Earth," in *Science and Civilisation in China*, vol. 3 (Cambridge: Cambridge University Press, 1959), 29.

48   J. J. O'Connor and E. F. Robertson, "Xu Yue," MacTutor (School of Mathematics and Statistics, University of St. Andrews, 2003), https://mathshistory.st-andrews.ac.uk/Biographies/Xu_Yue/; Needham and Wang, "Mathematics," 30.

49   Alexeï K. Volkov, "Large Numbers and Counting Rods," *Extrême Orient Extrême Occident* 16, no. 16 (1994): 71–92, https://doi.org/10.3406/oroc.1994.991.

50   Needham and Wang, "Mathematics," 70.

51   J. J. O'Connor and E. F. Robertson, "Chinese Numerals," MacTutor. School of Mathematics and Statistics, University of St. Andrews, 2004. https://mathshistory.st-andrews.ac.uk/HistTopics/Chinese_numerals/.

52   Lam Lay Yong and Ang Tian Se, *Fleeting Footsteps: Tracing the Concep-

tion of Arithmetic and Algebra in Ancient China, Revised (Singapore: World Scientific, 2004), 71–78, https://doi.org/10.1142/5425.

53   Needham and Wang, "Mathematics," 90, 115, 126–127.

54   Jean Claude Martzloff, *A History of Chinese Mathematics* (Berlin, Heidelberg: Springer, 2006), 209, https://doi.org/10.1007/978-3-540-33783-6.

55   Needham and Wang, "Mathematics," 8–9.

56   Martzloff, *History of Chinese Mathematics*, 210.

57   Gregory Blue, "Needham, (Noël) Joseph Terence Montgomery (1900–1995), Biochemist and Historian," *Oxford Dictionary of National Biography*, 2008, https://doi.org/10.1093/ref:odnb/58035.

58   Needham and Wang, "Mathematics," 77–78.

59   "Chinese Zhusuan, Knowledge and Practices of Mathematical Calculation Through the Abacus," Intangible Heritage (UNESCO, 2018), https://ich.unesco.org/en/RL/chinese-zhusuan-knowledge-and-practices-of -mathematical-calculation-through-the-abacus-00853.

60   Needham and Wang, "Mathematics," 51, 75–76; Menninger, *Number Words*, 310; William Aspray, ed., *Computing Before Computers* (Ames: Iowa State University Press, 1990), 14.

61   Martzloff, *History of Chinese Mathematics*, 216.

62   Menninger, *Number Words*, 310; Needham and Wang, "Mathematics," 74–75.

63   Needham and Wang, "Mathematics," 3–4.

64   Aspray, *Computing Before Computers*, 7–8.

65   Aspray, *Computing Before Computers*, 15.

66   Takashi Kojima, *Advanced Abacus: Japanese Theory and Practice* (Rutland, VT: C. E. Tuttle, 1963), 22; Wai-Ming Ng, "The Images of Yang Guifei in Tokugawa Texts," *Journal of Asian History* 50, no. 1 (April 22, 2016): 117–39, https://doi.org/10.13173/jasiahist.50.1.0117.

67   Kojima, *Advanced Abacus*, 23.

68   Mokoto Rich, "The Right Answer? 8,186,699,633,530,061 (An Abacus Makes It Look Almost Easy)," *New York Times*, August 21, 2019, https://www .nytimes.com/2019/08/21/world/asia/japan-abacus.html; Clyde Haberman, "In the Land of Sony, the Abacus Is Still King," *New York Times*, August 6, 1983, https://www.nytimes.com/1983/08/06/world/in-the-land-of-sony-the -abacus-is-still-king.html.

69   Menninger, *Number Words*, 307.

70   "Hands Down!," *Reader's Digest*, March 1947; Takashi Kojima, *The Japanese Abacus: Its Use and Theory* (Rutland, VT: C. E. Tuttle, 1954), 12.

71   "Hands Down!"; Menninger, *Number Words*, 307–309.

72   Rich, "The Right Answer?"; Haberman, "In the Land of Sony."

73   Menninger, *Number Words*, 305; Rudolf Fellmann, "Römische Rechentafeln Aus Bronze," *Antike Welt* 14, no. 1 (1983): 36–40.

74   Pullan, *History of the Abacus*, 19–20; J. J. O'Connor and E. F. Robertson, "Mathematics of the Incas," MacTutor (School of Mathematics and Statistics, University of St. Andrews, 2001).

75   Herbert Bruderer, "How Did the Romans Calculate?," BLOG@CACM (ACM, February 20, 2019), https://cacm.acm.org/blogs/blog-cacm/234881 -how-did-the-romans-calculate/fulltext.

76   Helmut Koenigsberger, "The End of the Ancient World, and the Beginning of the Middle Ages, 400–700," in *Medieval Europe: 400–1500* (Pearson Education, 1987), 9–66; Bernhard Bischoff, *Latin Palaeography: Antiquity and the Middle Ages* (Cambridge University Press, 1995), 190–201, http://www .worldcat.org/isbn/9780521367264.

77   Geoffrey Barraclough, "Holy Roman Empire," *Encyclopaedia Britannica* (Chicago: Encyclopaedia Britannica, August 12, 2013), http://www .britannica.com/EBchecked/topic/269851/Holy-Roman-Empire; Barrie Dobson and Michael Lapidge, "Carolingian Renaissance," *Encyclopedia of the Middle Ages* (Oxford University Press, 2000), http://www.oxfordreference.com/ view/10.1093/acref/9780227679319.001.0001/acref-9780227679319-e-483; G. W. Trompf, "The Concept of the Carolingian Renaissance," *Journal of the History of Ideas* 34, no. 1 (1973).

78   Menninger, *Number Words*, 322–325; Oscar G. Darlington, "Gerbert, the Teacher," *The American Historical Review* 52, no. 3 (April 1947): 456–476, https://doi.org/10.2307/1859882.

79   Menninger, *Number Words*, 322–325; Darlington, "Gerbert, the Teacher"; David Eugene Smith and Louis Charles Karpinski, *The Hindu-Arabic Numerals* (Boston: Ginn, 1911), 112–120.

80   Aspray, *Computing Before Computers*, 12.

81   Menninger, *Number Words*, 366–367.

82   Menninger, *Number Words*, 375–376.

83   Menninger, *Number Words*, 380–381.

84   Menninger, *Number Words*, 378.

85   Aspray, *Computing Before Computers*, 10; Stephen Johnston, "Recorde, Robert (c. 1512–1558), Mathematician," *Oxford Dictionary of National Biography*, 2008, https://doi.org/10.1093/ref:odnb/23241.

86   Barbara E. Reynolds, "The Algorists vs. the Abacists: An Ancient Controversy on the Use of Calculators," *The College Mathematics Journal* 24, no. 3 (April 24, 1993): 221, https://doi.org/10.2307/2686479.

87   Reynolds, "Algorists vs. Abacists, " 218.

88   J. J. O'Connor and E. F. Robertson, "Indian Numerals," MacTutor (School of Mathematics and Statistics, University of St. Andrews, 2000), https://mathshistory.st-andrews.ac.uk/HistTopics/Indian_numerals/; Reynolds, "Algorists vs. Abacists," 218.

89   "'Abbāsid Dynasty," *Encyclopaedia Britannica* (Chicago: Encyclopaedia Britannica, 2013), http://www.britannica.com/EBchecked/topic/613719/Umayyad-dynasty; J. J. O'Connor and E. F. Robertson, "Abu Ja'far Muhammad Ibn Musa Al-Khwarizmi," MacTutor (School of Mathematics and Statistics, University of St. Andrews, 1999), https://mathshistory.st-andrews.ac.uk/Biographies/Al-Khwarizmi/.

90   "Al-Khwārizmī," *Encyclopaedia Britannica*, accessed April 26, 2021, https://www.britannica.com/biography/al-Khwarizmi; O'Connor and Robertson, "Al-Khwarizmi."

91   Øystein Ore, *Number Theory and Its History* (New York: McGraw Hill, 1948), 20–21; Smith and Karpinski, *Hindu-Arabic Numerals*, 130–131.

92   Reynolds, "Algorists vs. Abacists," 218, 221; Smith and Karpinski, *Hindu-Arabic Numerals*, 133.

93   Reynolds, "Algorists vs. Abacists," 221.

94   Ore, *Number Theory*, 21.

95   Smith and Karpinski, *Hindu-Arabic Numerals*, 137–138.

3. THE SLIDE RULE

1   "Modernism," Tate (The Board of Trustees of the Tate Gallery), accessed June 1, 2020, https://www.tate.org.uk/art/art-terms/m/modernism.

2   Eugenia Russell, "Fall of Constantinople," *Oxford Bibliographies* (Oxford University Press, 2015), https://doi.org/10.1093/OBO/9780195399301-0192.

3   William Gilbert, "The Beginning of the Scientific Revolution," in *Renaissance and Reformation* (Lawrence, KS: Carrie, 1998).

4   Boyer, *History of Mathematics*, 333–342; J. J. O'Connor and E. F. Robertson, "Thomas Fincke," MacTutor (School of Mathematics and Statistics, University of St. Andrews, 2012), https://mathshistory.st-andrews.ac.uk/Biographies/Fincke/.

5   Boyer, *History of Mathematics*, 347–350; J. J. O'Connor and E. F. Robertson, "Simon Stevin," MacTutor (School of Mathematics and Statistics, University of St. Andrews, 2004), https://mathshistory.st-andrews.ac.uk/Biographies/Stevin/.

6   Boyer, *History of Mathematics*, 352; J. J. O'Connor and E. F. Robertson, "Ludolph Van Ceulen," MacTutor (School of Mathematics and Statistics, University of St. Andrews, 2009), https://mathshistory.st-andrews.ac.uk/Biographies/Van_Ceulen/.

7   Mark Napier, *Memoirs of John Napier of Merchiston, His Lineage, Life, and Times, with a History of the Invention of Logarithms* (Edinburgh: William Blackwood, 1834), 67.

8   Julian Goodare, "Napier, Sir Archibald, of Merchiston (1534–1608), Administrator," *Oxford Dictionary of National Biography* (Oxford University Press, 2004), https://doi.org/10.1093/ref:odnb/19744; George Molland, "Napier, John, of Merchiston (1550–1617), Mathematician," *Oxford Dictionary of National Biography* (Oxford University Press, 2010), https://doi.org/10.1093/ref:odnb/19758.

9   Goodare, "Napier, Sir Archibald"; Molland, "Napier, John."

10   Miloš Rydval et al., "Reconstructing 800 Years of Summer Temperatures in Scotland from Tree Rings," *Climate Dynamics* 49 (2017), https://doi.org/10.1007/s00382-016-3478-8.

11   Molland, "Napier, John."

12   Robert G. Clouse, "John Napier and Apocalyptic Thought," *The Sixteenth Century Journal* 5, no. 1 (1974): 101–114.

13   Clouse, "John Napier and Apocalyptic Thought."

14   Molland, "Napier, John."

15   Historic Environment Scotland, "Battle of Glenlivet (BTL33)," The Inventory of Historic Battlefields (Historic Environment Scotland, 2012), http://portal.historicenvironment.scot/designation/BTL33.

16   Hiram Morgan, "Teaching the Armada: An Introduction to the Anglo-Spanish War, 1585–1604," *History Ireland* 14, no. 5 (2006): 37–43.

17   Napier, *Memoirs of John Napier*, 246–248.

18   Napier, *Memoirs of John Napier*, 246–248.

19   Thomas Urquhart, *Ekskybalauron: Or, The Jewel*, Early English Books Online (London, 1652), 80–81.

20   R. D. S. Jack, "Urquhart [Urchard], Sir Thomas, of Cromarty (1611–1660), Author and Translator," *Oxford Dictionary of National Biography* (Oxford University Press, 2015), https://doi.org/10.1093/ref:odnb/28019; Denton Fox, "Admirable Urquhart," *London Review of Books* 6, no. 17 (1984), https://www.lrb.co.uk/the-paper/v06/n17/denton-fox/admirable-urquhart.

21   D. L. Simms, "Archimedes and the Burning Mirrors of Syracuse," *Technology and Culture* 18, no. 1 (1977): 1–24.

22   Napier, *Memoirs of John Napier*, 246–248.

23   Albert Kapr, *Johann Gutenberg : The Man and His Invention*, trans. Douglas Martin (Brookfield, VT: Scolar Press, 1996), 71–75.

24   Molland, "Napier, John."

25   Julian Goodare, "Between Humans and Angels: Scientific Uses for Fairies in Early Modern Scotland," in *Fairies, Demons and Nature Spirits. "Small Gods" at the Margins of Christendom.*, ed. Michael Ostling (Basingstoke: Palgrave MacMillan, 2017), 174; Goodare, "Napier, Sir Archibald."

26   Jonathan Andrews, "Napier, Richard (1559–1634), Astrological Physician and Church of England Clergyman," *Oxford Dictionary of National Biography* (Oxford University Press, 2009), https://doi.org/10.1093/ref:odnb/19763.

27   John Read, "Scottish Alchemy in the Seventeenth Century," *Chymia* 1 (1948): 142–145, https://doi.org/10.2307/27757120.

28   Molland, "Napier, John."

29   Lynn Thorndike, "The True Place of Astrology in the History of Science," *Isis* 46, no. 3 (June 28, 1955): 273–278; T. G. Cowling, "Mathematicians, Astrology and Cosmogony," *Quarterly Journal of the Royal Astronomical Society* 18 (June 1977): 199.

30   Peter Wright, "Astrology and Science in Seventeenth-Century England," *Social Studies of Science* 5, no. 4 (November 1975): 399–422.

31   Cowling, "Mathematicians, Astrology and Cosmogony."

32   H. J. Easterling, "Homocentric Spheres in 'De Caelo,'" *Phronesis* 6, no. 2 (1961): 138–153.

33   O. Neugebauer, "Mathematical Methods in Ancient Astronomy," *Bulletin of the American Mathematical Society* 54, no. 11, Part 1 (1948): 1013–1041,

https://projecteuclid.org:443/euclid.bams/1183512467; Thorndike, "True Place of Astrology," 1.

34   J. J. O'Connor and E. F. Robertson, "Hipparchus of Rhodes," MacTutor (School of Mathematics and Statistics, University of St. Andrews, 1999), https://mathshistory.st-andrews.ac.uk/Biographies/Hipparchus/; Neugebauer, "Mathematical Methods in Ancient Astronomy."

35   Boyer, *History of Mathematics*, 342.

36   Enrique A. González-Velasco, "Mathematical Introduction," in *The Life and Works of John Napier* (Cham: Springer, 2017), 393–394, https://doi.org/10.1007/978-3-319-53282-0; John Napier and Mark Napier, *De Arte Logistica* (Edinburgh: Bannatyne Club, 1839).

37   Victor E. Thoren, "Prosthaphaeresis Revisited," *Historia Mathematica* 15, no. 1 (1988): 32–39.

38   Boyer, *History of Mathematics*, 342; Molland, "Napier, John."

39   Florian Cajori, "History of the Exponential and Logarithmic Concepts," *The American Mathematical Monthly* 20, no. 1 (1913): 5–14.

40   "Algorithm, N.," OED Online, March 2021, https://www.oed.com/view/Entry/4959.

41   Aspray, *Computing Before Computers*, 24–25; Molland, "Napier, John."

42   Erwin Tomash and Michael R. Williams, "N Chapter," in *The Erwin Tomash Library on the History of Computing: An Annotated and Illustrated Catalog* (Erwin Tomash and Michael R. Williams., n.d.), http://www.cbi.umn.edu/hostedpublications/Tomash/index.htm; Boyer, *History of Mathematics*; Cajori, "Exponential and Logarithmic Concepts."

43   Robin J. Wilson, "The Gresham Professors of Geometry Part 1: The First One Hundred Years," *BSHM Bulletin: Journal of the British Society for the History of Mathematics* 32, no. 2 (2017): 125–135, https://doi.org/10.1080/17498430.2016.1236317; Boyer, *History of Mathematics*, 344–345.

44   Napier, *Memoirs of John Napier*, 409.

45   Wilson, "Gresham Professors of Geometry Part 1"; Denis Roegel, "A Reconstruction of Briggs' Logarithmorum Chilias Prima (1617)," 2010, https://hal.inria.fr/inria-00543935/.

46   Aspray, *Computing Before Computers*, 26–27.

47   Aspray, *Computing Before Computers*, 26.

48   B.O.B. Williams and R. G. Johnson, "Ready Reckoners," *IEEE Annals of*

the History of Computing 27, no. 4 (2005): 64, https://doi.org/10.1109/MAHC .2005.64.

49   Otto van Poelje, "Gunter Rules in Navigation," *Journal of the Ough-tred Society* 13 (2004): 11–22; J. J. O'Connor and E. F. Robertson, "Edmund Gunter," MacTutor (School of Mathematics and Statistics, University of St. Andrews, 2017), https://mathshistory.st-andrews.ac.uk/Biographies/Gunter/; Jim Bennett, "Early Modern Mathematical Instruments," *Isis* 102, no. 4 (2011): 697–705.

50   H. K. Higton, "Gunter, Edmund (1581–1626), Mathematician," *Oxford Dictionary of National Biography* (Oxford University Press, 2004), https://doi.org/10.1093/ref:odnb/11751; Aspray, *Computing Before Computers*, 27–28.

51   John Aubrey, *"Brief Lives," Chiefly of Contemporaries*, vol. 2 (Oxford: Clarendon Press, 1898), 214–215.

52   Higton, "Gunter, Edmund"; Wilson, "Gresham Professors of Geometry Part 1."

53   Wilson, "Gresham Professors of Geometry Part 1."

54   Edmund Gunter, *The Description and Use of the Sector, the Crosse-Staffe and Other Instruments* (London: Edmund Weaver, 1624).

55   Higton, "Gunter, Edmund"; Charles H. Cotter, "Edmund Gunter (1581–1626)," *Journal of Navigation* 34, no. 3 (1981): 363–367, https://doi .org/10.1017/S0373463300047998.

56   Michael R. Williams and Erwin Tomash, "The Sector: Its History, Scales, and Uses," *IEEE Annals of the History of Computing* 25, no. 1 (January 2003): 34–47, https://doi.org/10.1109/MAHC.2003.1179877.

57   Herbert Bruderer, "How Do You Calculate with the Sector?," BLOG@ CACM (Association for Computing Machinery, 2019), https://cacm.acm.org/ blogs/blog-cacm/237293-how-do-you-calculate-with-the-sector/; van Poelje, "Gunter Rules in Navigation"; C. J. Sangwin, "Edmund Gunter and the Sector," 2003.

58   van Poelje, "Gunter Rules in Navigation"; Sangwin, "Edmund Gunter and the Sector."

59   van Poelje, "Gunter Rules in Navigation"; Sangwin, "Edmund Gunter and the Sector."

60   van Poelje, "Gunter Rules in Navigation"; "Gunter's Scale Signed Mer-rifield & Co.," National Museum of American History (Washington, DC:

Smithsonian Institution), accessed July 27, 2020, https://americanhistory.si
.edu/collections/search/object/nmah_1214919.

61   Frances Willmoth, "Oughtred, William (Bap. 1575, d. 1660), Mathematician," *Oxford Dictionary of National Biography*, August 2008, https://doi
.org/10.1093/ref:odnb/20950; Aspray, *Computing Before Computers*.

62   Willmoth, "Oughtred, William"; Katherine Hill, " 'Juglers or Schollers?':
Negotiating the Role of a Mathematical Practitioner," *The British Journal for
the History of Science* 31, no. 3 (1998): 253–274.

63   Aspray, *Computing Before Computers*, 28–29; Hill, " 'Juglers or
Schollers?' "

64   Aspray, *Computing Before Computers*, 28–29; Florian Cajori, *On the History of Gunter's Scale and the Slide Rule During the 17th Century* (Berkeley:
University of California Press, 1920), 198.

65   Aspray, *Computing Before Computers*, 28–29.

66   Aspray, *Computing Before Computers*, 28–29.

67   Hill, " 'Juglers or Schollers?' "

68   Florian Cajori, *A History of the Logarithmic Slide Rule and Allied Instruments* (London: Constable, 1909), 16–17; Ted Hume and R. J. Koppany, *The
Oughtred Society Slide Rule Reference Manual* (The Oughtred Society, 2012),
35; A. F. Pollard and H. K. Higton, "Partridge, Seth (1603/4–1686), Mathematical Writer," *Oxford Dictionary of National Biography* (Oxford University
Press, 2004), https://doi.org/10.1093/ref:odnb/21490.

69   Cajori, *History of the Logarithmic Slide Rule*, 20.

70   Tom Martin, "Gauging: The Art behind the Slide Rule," *Brewery History*, no. 133 (2009): 69–86, http://breweryhistory.com/journal/archive/133/
Gauging.pdf.

71   Ron Manley, "Gauging," sliderules.info, accessed August 9, 2020, http://
www.sliderules.info/a-to-z/gauging.htm.

72   Cajori, *History of the Logarithmic Slide Rule*, 17–20.

73   Aspray, *Computing Before Computers*, 31–32; Nathan Zeldes, "The History of the Cursor and Evolution of the Slide Rule," *History of Computing*,
2012, https://www.nzeldes.com/HOC/Cursors.htm.

74   Cajori, *History of the Logarithmic Slide Rule*, 20; Martin, "Gauging";
Aspray, *Computing Before Computers*, 31–32.

75   John Rabone & Sons, *Price List of Boxwood, Ivory, Brass and Steel
Measuring Rules, Engine Divided Steel Straight Edges, Measuring Tapes, with*

Metallic Wire, Steel, & Linen Tapes, Brass, Iron, and Wood Spirit Levels, &c., Manufactured by John Rabone & Sons, (Established 1784) (Birmingham: Albert Steam Printing Works, 1878).

76   Arthur W. Clayden, "The Actinograph: An Instrument for Observing and Recording Changes in Radiation," *Quarterly Journal of the Royal Meteorological Society* 37, no. 158 (1911): 163–168, https://doi.org/10.1002/qj .49703715804.

77   Martin Brenner, "Pilot Balloon Slide Rule MK IV Detail," Martin Brenner's Pilot Balloon Resources, accessed August 12, 2020, https://web .csulb.edu/~mbrenner/mk_iv_detail.htm; "J. J. Hicks Pilot Balloon Slide Rule," National Museum of American History (Washington, DC: Smithsonian Institution), accessed August 13, 2020, https://americanhistory.si.edu/ collections/search/object/nmah_1215046.

78   Henry Aldinger and Ed Chamberlain, "Gilson Slide Rules—Part II— The Large Rules," *Journal of the Oughtred Society* 9, no. 2 (2000): 47–58.

79   "Fisher Control Valve Sizing Rule" (Rochester Avionic Archives), accessed August 12, 2020, https://rochesteravionicarchives.co.uk/collection/ general-purpose/slide-rule-2.

80   David Rance, "Disney (Walt Disney Productions): Mickey Mouse," David's calculating sticks, 2017, https://sliderules.nl/mysliderules/detail/ disney-mickey-mouse-disney-world-calcuator-coversion-aid.

81   Pablo Valerio, "E6B Computer: Celebrating 75 Years of Flight," *InformationWeek*, December 30, 2015, https://www.informationweek.com/ government/e6b-computer-celebrating-75-years-of-flight/a/d-id/1323695.

82   "The Ultimate Wrist-Worn Instrument for Pilots," Breitling, accessed August 15, 2020, https://www.breitling.com/gb-en/heritage/navitimer/; Isabella Proia, "Breitling Navitimer: The Early Years," Phillips, 2018, https:// www.phillips.com/article/37929545/breitling-navitimer-the-early-years.

83   Alex Green and Jesse Gordon, "Slide Rules and WWII Bombing: A Personal History," *Journal of the Oughtred Society* 21 (2012): 44–48.

84   "Circular Slide Rule, 'Emergency Computor', for Lancaster Bombsight Mark XIV," Science Museum Group, accessed August 14, 2020, https:// collection.sciencemuseumgroup.org.uk/objects/co60926/circular-slide-rule -emergency-computor-for-lan-circular-slide-rule-military.

85   Alex Green, "How Slide Rules Won a War," *Journal of the Oughtred Society* 14 (2005): 1–8.

86   B. W. Soole, "The RADIAC Slide Rule for the Computation of External Radiation Dose from Nuclear Fission Products," *Journal of Scientific Instruments* 29, no. 6 (June 1952): 189–192, https://doi.org/10.1088/0950-7671/29/6/308; Paul Frame, "Nuclear Slide Rules," Oak Ridge Associated Universities, accessed August 15, 2020, https://www.orau.org/ptp/collection/sliderules/sliderules.htm.

87   Robert A. James, "The Deaths of the Slide Rule," *Journal of the Oughtred Society* 23, no. 2 (2014): 6–17.

88   Michael Freudiger et al., "Mathematics on the Moon: The 'Apollo' Pickett," *Journal of the Oughtred Society* 10, no. 2 (2001): 15–18.

89   Freudiger et al., "Mathematics on the Moon."

90   "Slide Rule, 5-Inch, Pickett N600-ES, Apollo 13," National Air and Space Museum, accessed August 17, 2020, https://airandspace.si.edu/collection-objects/slide-rule-5-inch-pickett-n600-es-apollo-13/nasm_A19840160000.

91   "Pickett N600-ES"; Peter D. Kramer, "Moon Landing: No Small Step, No Giant Leap Without NJ Slide Rules," *NorthJersey*, December 10, 2019, https://eu.northjersey.com/story/news/2019/07/19/apollo-11-slide-rules/1742820001/.

92   Cliff Stoll, "When Slide Rules Ruled," *Scientific American* 294, no. 5 (2006): 80–87.

93   James, "Deaths of the Slide Rule."

4. THE CLOCKWORK AGE

1   Elizabeth King, "Clockwork Prayer: A Sixteenth-Century Mechanical Monk," *Blackbird* 1, no. 1 (2002), https://blackbird.vcu.edu/v1n1/nonfiction/king_e/prayer_introduction.htm; Clare Vincent and J. H. Leopold, "Seventeenth-Century European Watches," Heilbrunn Timeline of Art History (New York: The Metropolitan Museum of Art, March 2009), https://www.metmuseum.org/toah/hd/watc/hd_watc.htm; Clare Vincent, "Diana, Goddess of the Showering Arrows: A Seventeenth-Century German Automaton Clock," *Yale University Art Gallery Bulletin*, 2016, 75–87; Thomas Patteson, "'The Joy of Precision': Mechanical Instruments and the Aesthetics of Automation," in *Instruments for New Music: Sound, Technology, and Modernism* (Oakland, CA: University of California Press, 2016), 18–51.

2   René Descartes, *The Philosophical Writings of Descartes*, ed. John Cottingham, Robert Stoothoff, and Dugald Murdoch, vol. 1 (Cambridge: Cambridge University Press, 1985), https://doi.org/10.1017/CBO9780511805042.

3   Amelia Soth, "The Marvelous Automata of Antiquity," JSTOR Daily (ITHAKA, 2018), https://daily.jstor.org/the-marvelous-automata-of-antiquity/.

4   Archimedes and Donald R. Hill, *On the Construction of Water-Clocks: Kitab Arshimidas Fi `amal Al-Binkamat* (London: Turner & Devereux, 1976); J. G. Greenwood and B. Woodcroft, *The Pneumatics of Hero of Alexandria: From the Original Greek* (London: Taylor, Walton and Maberly, 1851), 72; G. Nadarajan, "Islamic Automation: A Reading of Al-Jazari's The Book of Knowledge of Ingenious Mechanical Devices (1206)," 2007.

5   Silvio A. Bedini, "The Role of Automata in the History of Technology," *Technology and Culture* 5, no. 1 (1964): 31–32.

6   Alan J. Friedman, "The Clockwork Universe," *Technology and Culture* 25, no. 2 (1984): 284; Bedini, "The Role of Automata," 32.

7   D. Masson et al., *Macmillan's Magazine*, vol. 10, English Literary Periodicals (London: Macmillan and Company, 1864), 438.

8   Thomas Hockey et al., eds., "Schickard, Wilhelm," *The Biographical Encyclopedia of Astronomers* (New York: Springer, 2007), https://doi.org/10.1007/978-0-387-30400-7_1233; J. J. O'Connor and E. F. Robertson, "Wilhelm Schickard," MacTutor (School of Mathematics and Statistics, University of St. Andrews, April 2009), https://mathshistory.st-andrews.ac.uk/Biographies/Schickard/.

9   Hockey et al., "Kepler, Johannes," *The Biographical Encyclopedia of Astronomers*; O'Connor and Robertson, "Wilhelm Schickard."

10   Hockey et al., "Schickard, Wilhelm," *The Biographical Encyclopedia of Astronomers*.

11   Aspray, *Computing Before Computers*, 35.

12   J.A.N. Lee, "Wilhelm Schickard," Computer Pioneers (Institute of Electrical and Electronics Engineers, 2013), https://history.computer.org/pioneers/schickard.html.

13   Lee, "Wilhelm Schickard."

14   Aspray, *Computing Before Computers*, 38.

15   Franz Hammer, "Nicht Pascal Sondern Der Tübinger Professor Wilhelm Schickard Erfand Die Rechenmaschine! 20 Jahre Vor Pascal Konstru-

ierte Schickard Schon Eine Vier-Spezies- Rechenmaschine," *Büromarkt*, no. 20 (1958): 1023–1025; Friedrich W. Kistermann, "How to Use the Schickard Calculator," *IEEE Annals of the History of Computing* 23, no. 1 (2001): 80–85, https://doi.org/10.1109/85.929917.

16 John Napier, *Rabdologiae* (Edinburgh: Andro Hart, 1617); Kistermann, "How to Use the Schickard Calculator"; Molland, "Napier, John."

17 Aspray, *Computing Before Computers*, 39; Lee, "Wilhelm Schickard."

18 Lee, "Wilhelm Schickard"; O'Connor and Robertson, "Wilhelm Schickard."

19 J. J. O'Connor and E. F. Robertson, "Blaise Pascal," MacTutor (School of Mathematics and Statistics, University of St. Andrews, 1996), https://mathshistory.st-andrews.ac.uk/Biographies/Pascal/; N. Wirth, "The Programming Language Pascal," *Acta Informatica* 1, no. 1 (March 1, 1971): 35–63, https://doi.org/10.1007/BF00264291; "2.3.4 Derived Units," The International System of Units (SI) (National Institute of Standards and Technology, March 10, 2021), https://www.nist.gov/pml/special-publication-330/sp-330 -section-2#2.3.4.

20 O'Connor and Robertson, "Blaise Pascal."

21 O'Connor and Robertson, "Blaise Pascal"; J. J. O'Connor and E. F. Robertson, "Marin Mersenne," MacTutor (School of Mathematics and Statistics, University of St. Andrews, 2005), https://mathshistory.st-andrews.ac.uk/ Biographies/Mersenne/.

22 O'Connor and Robertson; Chris K. Caldwell, "The Largest Known Primes," PrimePages, April 30, 2021, https://primes.utm.edu/largest.html; Divesh Aggarwal et al., "A New Public-Key Cryptosystem via Mersenne Numbers," in *Annual International Cryptology Conference* (Cham: Springer, 2018), 459–482.

23 O'Connor and Robertson, "Blaise Pascal."

24 Desmond Clarke, "Blaise Pascal," ed. Edward N Zalta, *The Stanford Encyclopedia of Philosophy* (Metaphysics Research Lab, Stanford University, 2015), https://plato.stanford.edu/archives/fall2015/entries/pascal/.

25 René Descartes, "8 Dec 1647: Descartes, René to Huygens, Constantijn," ed. Howard Hotson and Miranda Lewis, Early Modern Letters Online, accessed April 30, 2021, http://emlo.bodleian.ox.ac.uk/profile/work/6011566a -b16a-45bc-9a4d-bcd2dc12ab3b.

26 René Taton, "Sur l'invention de La Machine Arithmétique," *Revue d'his-*

toire Des Sciences et de Leurs Applications 16, no. 2 (1963): 139–160, https://doi.org/10.3406/rhs.1963.4447.

27 Blaise Pascal, *Les lettres de Blaise Pascal: accompagnées de lettres de ses correspondants*, ed. Maurice Beaufreton (Paris: G. Crès, 1922), 181–185, http://gallica.bnf.fr/ark:/12148/bpt6k69975r.

28 Taton, "Sur l'invention de La Machine Arithmétique."

29 Taton, "Sur l'invention de La Machine Arithmétique."

30 Aspray, *Computing Before Computers*, 40.

31 Taton, "Sur l'invention de La Machine Arithmétique."

32 S. Chapman, "Blaise Pascal (1623–1662): Tercentenary of the Calculating Machine," *Nature* 150, no. 3809 (1942): 509, https://doi.org/10.1038/150508a0.

33 Jean-François Gauvin, *Habits of Knowledge: Artisans, Savants and Mechanical Devices in Seventeenth-Century French Natural Philosophy* (Cambridge, MA: Harvard University, 2008), 230; "Huntsman Automaton (WB.134)," The Waddesdon Bequest (British Museum), accessed May 4, 2021, http://wb.britishmuseum.org/MCN2560#1500443001.

34 Taton, "Sur l'invention de La Machine Arithmétique."

35 Pascal, *Les lettres*, 181–184; Taton, "Sur l'invention de La Machine Arithmétique."

36 Clarke, "Blaise Pascal."

37 O'Connor and Robertson, "Blaise Pascal."

38 Clarke, "Blaise Pascal."

39 David Simpson, "Pascal, Blaise," *Internet Encyclopedia of Philosophy*, accessed May 4, 2021, https://iep.utm.edu/pascal-b/#SH2b.

40 Alan Marshall, "Morland, Sir Samuel, First Baronet (1625–1695), Natural Philosopher and Diplomat," Oxford Dictionary of National Biography, 2008, https://doi.org/10.1093/ref:odnb/19282; "The History of Magdalene College" (Cambridge: Magdalene College), accessed May 6, 2021, https://www.magd.cam.ac.uk/about/history.

41 Marshall, "Morland, Sir Samuel."

42 John Morrill, "Cromwell, Oliver (1599–1658), Lord Protector of England, Scotland, and Ireland," *Oxford Dictionary of National Biography*, 2015, https://doi.org/10.1093/ref:odnb/6765; Ruth Kleinman, "Belated Crusaders: Religious Fears in Anglo-French Diplomacy, 1654–1655," *Church History* 44, no. 1 (May 6, 1975): 34–46, https://doi.org/10.2307/3165097; George H. Clark, "Foreign Policy," in *Oliver Cromwell*

(New York: Harper & Bros., 1895), 157–182, http://www.worldcat.org/oclc/360422.

43 Morrill, "Cromwell, Oliver."

44 F. Cajori, "Rahn's Algebraic Symbols," *The American Mathematical Monthly* 31, no. 2 (1924): 65–71. F. Cajori, "Signs for Division and Ratio" (Cosimo, 2011), 268–278.

45 Samuel Morland, *The History of the Evangelical Churches of the Valleys of Piemont: Containing a Most Exact Geographical Description of the Place, and a Faithfull Account of the Doctrine, Life, and Persecutions of the Ancient Inhabitants* (London: Henry Hills, for Adoniram Byfield, 1658).

46 H. F. McMains, *The Death of Oliver Cromwell* (Lexington: University Press of Kentucky, 2000), 75.

47 Samuel Pepys et al., *The Diary of Samuel Pepys*, vol. 1 (London: George Bell & Sons, 1893), 137–138; A. Marshall, *Sir Samuel Morland and Stuart Espionage* (Tunbridge Wells: Friends of King Charles the Martyr Church, 2003), 10, http://researchspace.bathspa.ac.uk/3185/.

48 Morrill, "Cromwell, Oliver."

49 John Willcock, *Life of Sir Henry Vane the Younger, Statesman & Mystic (1613–1662)* (London: The Saint Catherine Press, 1913), 376.

50 Samuel Morland, *The Description and Use of Two Arithmetick Instruments: Together with a Short Treatise, Explaining and Demonstrating the Ordinary Operations of Arithmetick* (London: Moses Pitt, 1673).

51 Morland, *Description and Use of Two Arithmetick Instruments*; J. R. Ratcliff, "Samuel Morland and His Calculating Machines c.1666: The Early Career of a Courtier-Inventor in Restoration London," *The British Journal for the History of Science* 40, no. 2 (May 6, 2007): 170–172.

52 Ratcliff, "Samuel Moreland and His Calculating Machines," 178.

53 Ratcliff, "Samuel Morland and His Calculating Machines," 168–170.

54 Ratcliff, "Samuel Morland and His Calculating Machines," 176–178.

55 Jeremy Norman, "Samuel Morland Writes the First Book on a Calculating Machine Published in English," History of Information, accessed May 7, 2021, https://www.historyofinformation.com/detail.php?id=393.

5. THE ARITHMOMETER AND THE CURTA

1 "Calculating Machine," *The Manufacturer and Builder : A Practical Journal of Industrial Progress.* (New York: Western & Company, January 1872).

2 Stephen Johnston, "Making the Arithmometer Count," *Bulletin of the Scientific Instrument Society*, 1997, 12–21, https://www.mhs.ox.ac.uk/staff/saj/arithmometer/.

3 "Alsace," *Encyclopaedia Britannica* (Chicago: Encyclopaedia Britannica, n.d.), https://www.britannica.com/place/Alsace; J. Joly, "Un Grand Inventeur Alsacien: Charles-Xavier Thomas," in *La Vie En Alsace* (Strasbourg, 1932), 129–136.

4 Joly, "Grand Inventeur Alsacien," 130.

5 Joly, "Grand Inventeur Alsacien," 131–134.

6 Charles Xavier Thomas, Machine appelée arithmomètre, propre à suppléer à la mémoire et à l'intelligence dans toutes les opérations d'arithmétique, 1BA1447 (France, issued 1820), http://bases-brevets19e.inpi.fr/Thot/FrmFicheDoc.asp?idfiche=0021076&refFiche=0020758.

7 Johnston, "Making the Arithmometer Count."

8 *The Encyclopaedia Britannica*, vol. XVI Leibnitz (Leibniz), Gottfried Wilhem; Brandon C. Look, "Gottfried Wilhelm Leibniz," in *The Stanford Encyclopedia of Philosophy*, ed. Edward N Zalta, Spring 2020 (Metaphysics Research Lab, Stanford University, 2020), https://plato.stanford.edu/archives/spr2020/entries/leibniz/.

9 James A. Ryan, "Leibniz' Binary System and Shao Yong's 'Yijing,'" *Philosophy East and West* 46, no. 1 (1996): 59–90.

10 J. J. O'Connor and E. F. Robertson, "Gottfried Wilhelm von Leibniz," MacTutor (School of Mathematics and Statistics, University of St. Andrews, 1998), https://mathshistory.st-andrews.ac.uk/Biographies/Leibniz/.

11 "Elector (German Prince)," *Encyclopaedia Britannica*, 2013, https://www.britannica.com/topic/elector; Look, "Leibniz."

12 Look, "Leibniz."

13 Florin Stefan Morar, "Reinventing Machines: The Transmission History of the Leibniz Calculator," *British Journal for the History of Science* 48, no. 1 (March 6, 2015): 4, https://doi.org/10.1017/S0007087414000429; Ratcliff, "Samuel Morland and His Calculating Machines," 178.

14 Morar, "Reinventing Machines," 4.

15 Aspray, *Computing Before Computers*, 42–44; Morar, "Reinventing Machines," 18.

16 F. W. Kistermann, "When Could Anyone Have Seen Leibniz's Stepped Wheel?," *IEEE Annals of the History of Computing* 21, no. 2 (1999): 68–72,

https://doi.org/10.1109/85.761796; Johnston, "Making the Arithmometer Count."

17   L. Null and J. Lobur, *The Essentials of Computer Organization and Architecture* (Sudbury, MA: Jones and Bartlett Publishers, 2006), 56–57.

18   Johnston, "Making the Arithmometer Count."

19   René Tresse, "Le Conservatoire Des Arts et Métiers et La Société d'Encouragement Pour l'Industrie Nationale Au Début Du XIXe Siècle," *Revue d'histoire Des Sciences et de Leurs Applications* 5, no. 3 (1952), https://doi.org/10.3406/rhs.1952.2946; Francœur, "Rapport Fait Par M. Francœur, Au Nom Du Comité Des Arts Mécaniques, Sur La Machine à Calculer de M. Le Chevalier Thomas, de Colmar," *Bulletin de La Société d'encouragement Pour l'industrie Nationale*, vol. 212 (Paris, 1822).

20   Joly, "Grand Inventeur Alsacien," 131.

21   Johnston, "Making the Arithmometer Count."

22   Arthur Chandler, "L'Exposition Publique Des Produits de l'industrie Française," accessed May 14, 2021, http://www.arthurchandler.com/1798-exposition.

23   Johnston, "Making the Arithmometer Count."

24   Timo Leipälä, "The Life and Works of W. T. Odhner, Part I," in *2. Greifswalder Symposium Zur Entwicklung Der Rechentechnik* (Greifswald: Univ. Greifswald, Inst. für Mathematik und Informatik, 2003), 10.

25   R.C.A., "P. G. Scheutz, Publicist, Author, Scientific Mechanician, and Edvard Scheutz, Engineer—Biography and Bibliography," *Mathematical Tables and Other Aids to Computation* 2, no. 18 (1947): 238–245; Johnston, "Making the Arithmometer Count."

26   Michel Arnold, "Thomas de Colmar, Propriétaire Du Château de Maisons de 1850 à 1870," *Le Bulletin de La Société Des Amis Du Château de Maisons*, no. 1 (2006); Johnston, "Making the Arithmometer Count."

27   Ernst Martin, Peggy Aldrich Kidwell, and Michael R. Williams, *The Calculating Machines: Their History and Development* (Cambridge, MA: MIT Press, 1992), 53–54.

28   Johnston, "Making the Arithmometer Count"; "Currency Exchange Rates," XE.com, 2021, https://xe.com/; "Inflation Calculator" (Bank of England, 2021), https://www.bankofengland.co.uk/monetary-policy/inflation/inflation-calculator.

29   Martin Campbell-Kelly, "Large-Scale Data Processing in the Prudential,

1850–1930," *Accounting, Business & Financial History* 2, no. 2 (1992): 123, https://doi.org/10.1080/09585209200000036.

30   Erwin Tomash, "An Interview with Curt Herzstark (OH 140)," Charles Babbage Institute, 1987, 4–7.

31   Tomash, "Interview with Curt Herzstark," 10–11; "Adding Machines—Ten Keys & Fewer," National Museum of American History (Washington, DC: Smithsonian Institution), accessed May 17, 2021, https://americanhistory.si.edu/collections/search/object/nmah_1215046.

32   Tomash, "Interview with Curt Herzstark," 23.

33   Cliff Stoll, "The Curious History of the First Pocket Calculator," *Scientific American* 290, no. 1 (2004): 92–99; "Weight-for-Age" (World Health Organisation), accessed May 17, 2021, https://www.who.int/tools/child-growth-standards/standards/weight-for-age.

34   Tomash, "Interview with Curt Herzstark," 23–26, 40.

35   Michael Gehler, "Schuschnigg, Kurt," Deutsche Biographie 23 (Bayerische Staatsbibliothek, 2007), https://www.deutsche-biographie.de/pnd118762702.html; Katharina Ziegler, "Anschluss," AEIOU Österreich-Lexikon (Graz: Austria-Forum, March 15, 2021), https://austria-forum.org/af/AEIOU/Anschluss.

36   Tomash, "Interview with Curt Herzstark," 22, 26.

37   Tomash, "Interview with Curt Herzstark," 27.

38   Tomash, "Interview with Curt Herzstark," 29; "Buchenwald Concentration Camp, 1937–1945," Buchenwald Memorial (Buchenwald and Mittelbau-Dora Memorials Foundation), accessed May 18, 2021, https://www.buchenwald.de/en/72/.

39   Tomash, "Interview with Curt Herzstark," 31–32.

40   Tomash, "Interview with Curt Herzstark," 36–37; "1945—After the Liberation," Buchenwald Memorial (Buchenwald and Mittelbau-Dora Memorials Foundation), accessed May 18, 2021, https://www.buchenwald.de/en/464/.

41   William Gibson, "Math Grenades," in *Pattern Recognition* (London: Penguin, 2004); Donald E. Morse, "Advertising and Calculators in William Gibson's 'Pattern Recognition,'" *Science Fiction Studies* 31, no. 2 (2004): 330–332.

42   "Curta Type I Calculating Machine (DW0973)," Collection of Historical Scientific Instruments (Cambridge, MA: Harvard University, 1964),

http://waywiser.fas.harvard.edu/objects/3778/; Tomash, "Interview with Curt Herzstark," 63.

43   "Leaflet, Your CURTA Calculator," National Museum of American History (Washington, DC: Smithsonian Institution), accessed May 18, 2021, https://www.si.edu/es/object/nmah_904705.

44   Tomash, "Interview with Curt Herzstark," 24.

45   "Leaflet, Your CURTA Calculator."

46   "Miniature Machine Performs Engineering Calculations," *Product Engineering*, no. 23 (October 1952): 160–161, http://www.vcalc.net/cu-pe.htm.

47   Tomash, "Interview with Curt Herzstark," 36.

48   Tomash, "Interview with Curt Herzstark," 38–41.

49   Tomash, "Interview with Curt Herzstark," 41–43.

50   Tomash, "Interview with Curt Herzstark," 43–44.

51   Stoll, "First Pocket Calculator."

52   Tomash, "Interview with Curt Herzstark," 46–51, 55, 64; Stoll, "First Pocket Calculator."

53   Tomash, "Interview with Curt Herzstark," 51.

54   Stoll, "The First Pocket Calculator."

55   Bruce Flamm, "The Amazing Curta," Vintage Calculators Web Museum, 1997, http://www.vintagecalculators.com/html/the_amazing_curta.html.

56   Tomash, "Interview with Curt Herzstark," 52.

57   Tomash, "Interview with Curt Herzstark," 52–54.

58   Stoll, "First Pocket Calculator"; Flamm, "The Amazing Curta."

59   "Curta Portable Calculator (Advertisement)," *Popular Mechanics*, May 1952; "CPI Inflation Calculator" (U.S. Bureau of Labor Statistics), accessed May 19, 2021, https://www.bls.gov/data/inflation_calculator.htm.

60   Tomash, "Interview with Curt Herzstark," 59–60.

61   "Arithmometer," History Computer, accessed May 19, 2021, https://history-computer.com/arithmometer-history-of-the-arithmometer-of-thomas-de-colmar/.

62   Stoll, "First Pocket Calculator"; Jim Bianchi, "The Curta Mechanical Calculator," *Rallye Mazagine*, March 1976.

63   Herbert Bruderer, "The World's Smallest Mechanical Parallel Calculator: Discovery of Original Drawings and Patent Documents from the 1950s in Switzerland," in *International Communities of Invention and Innovation*, ed. Arthur Tatnall and Christopher Leslie (Cham: Springer, 2016), 186–192.

6. THE FRIDEN STW-10

1   Erwin Danneels, "Trying to Become a Different Type of Company: Dynamic Capability at Smith Corona," *Strategic Management Journal* 32, no. 1 (January 1, 2011): 1–31, https://doi.org/10.1002/smj.863; "Company History" (Unisys), accessed May 28, 2021, https://www.unisys.com/aboutus/company-history; Frank Stephen Baldwin, "An Interview with the Father of the Calculating Machine" (Monroe Calculating Machine Company, 1919).

2   Martin, Kidwell, and Williams, *The Calculating Machines*, 23.

3   Robert Lewis, "Bletchley Park," *Encyclopaedia Britannica*, 2016, https://www.britannica.com/place/Bletchley-Park; Andrew Hodges, "Turing, Alan Mathison (1912–1954), Mathematician and Computer Scientist," *Oxford Dictionary of National Biography*, 2017, https://doi.org/10.1093/ref:odnb/36578.

4   Alan M. Turing, "On Computable Numbers, with an Application to the Entscheidungsproblem," *Proceedings of the London Mathematical Society* s2-42, no. 1 (1937): 230–265, https://doi.org/10.1112/plms/s2-42.1.230; Alan M. Turing, "Computing Machinery and Intelligence," *Mind* LIX, no. 236 (1950): 433–460, https://doi.org/10.1093/mind/LIX.236.433.

5   Roland Pease, "Alan Turing: Inquest's Suicide Verdict 'Not Supportable,'" BBC News, June 26, 2012, https://www.bbc.co.uk/news/science-environment-18561092; "Royal Pardon for Codebreaker Alan Turing," BBC News, December 24, 2013, https://www.bbc.co.uk/news/technology-25495315.

6   Turing, "Computing Machinery and Intelligence," 436–437.

7   Menninger, *Number Words*, 212, 306.

8   Heinrich Schreiber, *Ayn New Kunstlich Buech, Welches Gar Gewiß Vnd Behend Lernet Nach Der Gemainen Regel Detre, Welschen Practic, Regeln Falsi vñ Etlichē Regeln Cosse Mancherlay Schöne Uñ Zu Wissen Notürfftig Rechnũg Auff Kauffmanschafft . . .* (Nürnberg: Stuchs, 1518), 13.

9   David Alan Grier, *When Computers Were Human* (Princeton: Princeton University Press, 2005), 13–14.

10   Z. E. Musielak and Billy Quarles, "The Three-Body Problem," *Reports on Progress in Physics* 77, no. 6 (2014): 65901.

11   J. J. O'Connor and E. F. Robertson, "Alexis Claude Clairaut," MacTutor (School of Mathematics and Statistics, University of St. Andrews, 1999), https://mathshistory.st-andrews.ac.uk/Biographies/Clairaut/.

12 Curtis Wilson, "Clairaut's Calculation of the Eighteenth-Century Return of Halley's Comet," *Journal for the History of Astronomy* 24, no. 1–2 (1993): 1–15, https://doi.org/10.1177/002182869302400101.

13 O'Connor and Robertson, "Clairaut."

14 Charles Bossut, *Histoire générale des mathématiques, depuis leur origine jusqu'à l'année 1808*, vol. 2 (Paris: F. Louis, 1810), 428–429; O'Connor and Robertson, "Clairaut."

15 Grier, *When Computers Were Human*, 55–56.

16 Steven J. Dick, "A History of the American Nautical Almanac Office," in *Nautical Almanac Office Sesquicentennial Symposium* (Washington, DC: Naval Observatory, 1999), 11–13, https://ui.adsabs.harvard.edu/abs/1999naos .symp...11D/abstract.

17 Dick, "American Nautical Almanac Office," 13.

18 Grier, *When Computers Were Human*, 66–67.

19 Grier, *When Computers Were Human*, 66–67.

20 Grier, *When Computers Were Human*, 50–52.

21 David Alan Grier, "The Human Computer and the Birth of the Information Age," in *70th Joseph Henry Lecture* (Washington, DC: PSW Science, 2001), https://pswscience.org/meeting/the-human-computer-and-the-birth-of -the-information-age/.

22 Grier, *When Computers Were Human*, 61.

23 Grier, "Human Computer and the Birth of the Information Age."

24 Grier, *When Computers Were Human*, 62.

25 Richard Holmes, "Maria Mitchell at 200: A Pioneering Astronomer Who Fought for Women in Science," *Nature* 558, no. 7710 (June 2018): 370–371, https://doi.org/10.1038/d41586-018-05458-6.

26 Grier, *When Computers Were Human*, 82–83.

27 Grier, "Human Computer and the Birth of the Information Age."

28 Grier, *When Computers Were Human*, 140–142.

29 Grier, *When Computers Were Human*, 146–147.

30 Grier, "Human Computer and the Birth of the Information Age."

31 David Alan Grier, "The Math Tables Project of the Work Projects Administration: The Reluctant Start of the Computing Era," *IEEE Annals of the History of Computing* 20, no. 3 (July 1998): 33, https://doi.org/10.1109/85 .707573.

32 J. J. O'Connor and E. F. Robertson, "Gertrude Blanch," MacTutor

(School of Mathematics and Statistics, University of St. Andrews, 2009), https://mathshistory.st-andrews.ac.uk/Biographies/Blanch/; David Alan Grier, "Gertrude Blanch of the Mathematical Tables Project," *IEEE Annals of the History of Computing* 19, no. 04 (October 1997): 20, https://doi.org/10.1109/85.627896.

33   Grier, *When Computers Were Human*, 212.

34   Grier, "Gertrude Blanch," 18; Grier, "Math Tables Project of the WPA," 45–46.

35   Grier, *When Computers Were Human*, 213.

36   Grier, "Gertrude Blanch," 20–21.

37   Grier, *When Computers Were Human*, 220.

38   Grier, *When Computers Were Human*, 218.

39   Jennifer S. Light, "When Computers Were Women," *Technology and Culture* 40, no. 3 (1999): 459.

40   Light, "When Computers Were Women, " 460–461.

41   Paul E. Ceruzzi, "When Computers Were Human," *IEEE Annals of the History of Computing* 13, no. 3 (July 1991): 243, https://doi.org/10.1109/MAHC.1991.10025.

42   Charles Frank, "Powell, Cecil Frank (1903–1969), Physicist," Oxford Dictionary of National Biography, 2004, https://doi.org/10.1093/ref:odnb/35588; Frederick Charles Frank and Donald Hill Perkins, "Cecil Frank Powell, 1903–1969," *Biographical Memoirs of Fellows of the Royal Society* 17 (November 30, 1971): 546, https://doi.org/10.1098/rsbm.1971.0021.

43   Grier, *When Computers Were Human*, 276.

44   Ceruzzi, "When Computers Were Human," 242–243.

45   Ted H. Skopinski and Katherine G. Johnson, *Determination of Azimuth Angle at Burnout for Placing a Satellite over a Selected Earth Position*, vol. 233 (National Aeronautics and Space Administration, 1960); Margot Lee Shetterly, *Hidden Figures: The Untold Story of the African American Women Who Helped Win the Space Race* (London: William Collins, 2017), 120, 219–220.

46   Shetterly, *Hidden Figures*, 215–217; Stephen J. Garber, "Friendship 7," NASA History (National Aeronautics and Space Administration, February 22, 2010), https://history.nasa.gov/friendship7/.

47   Rick Bensene, "Friden Model STW-10 Electro-Mechanical Calculator," The Old Calculator Web Museum, November 28, 1999, https://www.oldcalculatormuseum.com/fridenstw.html.

48 John Wolff, "Friden," John Wolff's Web Museum, May 20, 2020, http://www.johnwolff.id.au/calculators/Friden/Friden.htm.

49 John Wolff, "Calculators by Hans W. Egli, Zurich," John Wolff's Web Museum, March 14, 2016, http://www.johnwolff.id.au/calculators/Egli/Egli.htm.

50 Grier, *When Computers Were Human*, 243.

51 "Friden Fully Automatic Calculator Model STW Instruction Manual" (Rochester, NY: Friden Educational Center, 1963), ii.

52 John Wolff, "Full-Keyboard Rotary Calculators," John Wolff's Web Museum, December 8, 2017, http://www.johnwolff.id.au/calculators/fullcalc/fullcalc.htm.

53 Shetterly, *Hidden Figures*, 223.

54 Shetterly, *Hidden Figures*, 222.

55 John Wolff, "Pin-Wheel Calculators," John Wolff's Web Museum, November 10, 2020, http://www.johnwolff.id.au/calculators/pinwheel/pinwheel.htm.

56 Martin, Kidwell, and Williams, *Calculating Machines*, 89–94.

57 Joseph D. Zund, "Von Neumann, John Louis (1903–1957), Mathematician, Mathematical Physicist, and Theoretical Economist," American National Biography, 2000, https://doi.org/10.1093/anb/9780198606697.article.1301728.

58 Grier, *When Computers Were Human*, 295–297.

59 T. R. Kennedy Jr., "Electronic Computer Flashes Answers, May Speed Engineering," *New York Times*, February 15, 1946, https://www.nytimes.com/1946/02/15/archives/electronic-computer-flashes-answers-may-speed-engineering-new.html; "CPI Inflation Calculator."

60 "7090 Data Processing System," IBM Archives (IBM, October 4, 1960), https://www.ibm.com/ibm/history/exhibits/mainframe/mainframe_PP7090.html; "CPI Inflation Calculator."

7. THE CASIO 14-A

1 "Establishment of CASIO," Corporate History (CASIO), accessed June 9, 2021, https://world.casio.com/corporate/history/.

2 "Establishment of CASIO."

3 Joshua Hammer, *Yokohama Burning: The Deadly 1923 Earthquake*

and Fire That Helped Forge the Path to World War II (New York: Free Press, 2006), xiv.

4   Sam Roberts, "Kazuo Kashio, a Founder of Casio Computer, Dies at 89," *New York Times*, June 20, 2018, https://www.nytimes.com/2018/06/20/ obituaries/kazuo-kashio-founder-casio-computer-dies-at-89.html.

5   "Tadao Kashio Biography: History of Casio Computer Company," Astrum People, May 13, 2015, https://astrumpeople.com/tadao-kashio -biography/.

6   "Establishment of CASIO"; Terry McCarthy, "Obituary: Tadao Kashio," *The Independent*, October 23, 2011, https://www.independent.co.uk/news/ people/obituary-tadao-kashio-1496820.html.

7   "Establishment of CASIO."

8   "Establishment of CASIO"; "Tadao Kashio Biography."

9   "Hands Down!"; Kojima, *The Japanese Abacus*, 12; "樫尾俊雄について (About Toshio Kashio)," Toshio Kashio Memorial Museum of Invention, accessed June 10, 2021, https://kashiotoshio.org/about/.

10   Katsunori Kadokura, "Chronology of the History of Japanese Calculating Machines," X-Number, April 2000, http://www.xnumber.com/xnumber/ jmc_history.htm.

11   "About Toshio Kashio"; "Establishment of CASIO."

12   "Establishment of CASIO"; *A Dictionary of Physics* (Oxford: Oxford University Press, 2009), solenoid, https://doi.org/10.1093/acref/9780199233991 .001.0001.

13   Descriptions of the Kashios' early calculators are sparse. However, it is likely that they followed the same operating principles used in early electromechanical computers. See, for example, Claude E. Shannon, "A Symbolic Analysis of Relay and Switching Circuits," *Transactions of the American Institute of Electrical Engineers* 57, no. 12 (December 1938): 713–723, https://doi.org/10 .1109/T-AIEE.1938.5057767.

14   "Company Information" (Bunshodo Company Limited, 2017), http:// www.bun-sho-do.co.jp/english/corporate/about/; "Establishment of CASIO."

15   Andy Raskin, "Casio's Eccentric Product Culture, Built on Embracing Failure," Firm Narrative (Medium, 2018), https://medium.com/firm -narrative/some-say-they-embrace-failure-then-theres-casio-c5213315994c.

16   "Establishment of CASIO."

17   "Telegraph, N.," OED Online, December 2016, https://www.oed.com/view/Entry/198686.

18   Charles Ffoulkes, "Notes on the Development of Signals Used for Military Purposes," *Journal of the Society for Army Historical Research* 22, no. 85 (1943): 20–27; "July 20, 1805," *Journals of the Lewis and Clark Expedition* (Center for Digital Research in the Humanities), accessed June 11, 2021, https://lewisandclarkjournals.unl.edu/item/lc.jrn.1805-07-20; Ray Kerkhove, "Aboriginal Smoke Signalling and Signalling Hills in Resistance Warfare" (Sovereign Union of First Nations and Peoples in Australia, October 15, 2015), http://nationalunitygovernment.org/content/aboriginal-smoke-signalling-and-signalling-hills-resistance-warfare; Wenwu Chen et al., "Architectural Forms and Distribution Characteristics of Beacon Towers of the Ming Great Wall in Qinghai Province," *Journal of Asian Architecture and Building Engineering* 16, no. 3 (2017): 503–510, https://doi.org/10.3130/jaabe.16.503.

19   Alexander J. Field, "French Optical Telegraphy, 1793–1855: Hardware, Software, Administration," *Technology and Culture* 35, no. 2 (1994): 315–347.

20   E. A. Marland, "British and American Contributions to Electrical Communications," *The British Journal for the History of Science* 1, no. 1 (1962): 33.

21   Bernard S. Finn, "Morse, Samuel Finley Breese (1791–1872), Artist and Telegraph Inventor," American National Biography, 2000, https://doi.org/10.1093/anb/9780198606697.article.1301183.

22   David Hochfelder, "Joseph Henry: Inventor of the Telegraph?," The Joseph Henry Papers Project (Smithsonian Institution), accessed June 11, 2021, http://siarchives.si.edu/oldsite/siarchives-old/history/jhp/joseph20.htm; Wiliam O. Gibberd, "Davy, Edward (1806–1885)," *Australian Dictionary of Biography* (National Centre of Biography, Australian National University), accessed June 11, 2021, https://adb.anu.edu.au/biography/davy-edward-1966.

23   Lewis Coe, *The Telegraph: A History of Morse's Invention and Its Predecessors in the United States* (Jefferson, NC: McFarland, 1993), 30–31.

24   R. W. Burns, "Bell, Alexander Graham (1847–1922), Teacher of Deaf People and Inventor of the Telephone," *Oxford Dictionary of National Biography*, 2011, https://doi.org/10.1093/ref:odnb/30680.

25   G.W.O. Howe, "Alexander Graham Bell and the Invention of the Telephone," *Nature* 159, no. 4040 (1947): 455–457, https://doi.org/10.1038/159455a0.

26   Paul E. Ceruzzi, *Reckoners: The Prehistory of the Digital Computer, from Relays to the Stored Program Concept, 1935–1945* (Westport, CT: Greenwood Press, 1983), 74.

27   Burns, "Bell, Alexander Graham"; Howe, "Alexander Graham Bell"; Benjamin Lathrop Brown, "The Bell Versus Gray Telephone Dispute: Resolving a 144-Year-Old Controversy," *Proceedings of the IEEE* 108, no. 11 (November 1, 2020): 2083–2096, https://doi.org/10.1109/JPROC.2020.3017876.

28   "Bell Telephone Laboratories," Physics History Network (American Institute of Physics), accessed June 14, 2021, https://history.aip.org/phn/21506003.html.

29   J.A.N. Lee, "George Robert Stibitz," Computer Pioneers (Institute of Electrical and Electronics Engineers, 2012), https://history.computer.org/pioneers/stibitz.html; George R. Stibitz and E. Loveday, "Relay Computers at Bell-Labs," *Datamation* 13, no. 4 (1967): 35.

30   Eric W. Weisstein, "Complex Number," MathWorld (Wolfram Research, Inc.), accessed June 14, 2021, https://mathworld.wolfram.com/ComplexNumber.html.

31   E. D. Solomentsev, "Complex Number," *Encyclopedia of Mathematics* (European Mathematical Society), accessed June 14, 2021, https://encyclopediaofmath.org/index.php?title=Complex_number&oldid=18011.

32   Ceruzzi, *Reckoners*, 78.

33   Ceruzzi, *Reckoners*, 86–87.

34   Lee, "George Robert Stibitz"; J. Daintith, "teletypewriter," *Oxford Dictionary of Computing*, Oxford Paperback Reference (Oxford University Press, 2004); R. A. Nelson and K. M. Lovitt, *History of Teletypewriter Development* (Skokie, IL: Teletype Corporation, 1963).

35   Ronald Kline, "Shannon, Claude E. (30 Apr. 1916–24 Feb. 2001), Mathematician," American National Biography, 2018, https://doi.org/10.1093/anb/9780198606697.013.1302696.

36   Shannon, "Symbolic Analysis of Relay and Switching Circuits."

37   J. J. O'Connor and E. F. Robertson, "George Boole," MacTutor (School of Mathematics and Statistics, University of St. Andrews, 2004), https://mathshistory.st-andrews.ac.uk/Biographies/Boole/.

38   Ryan, "Leibniz' Binary System and Shao Yong's 'Yijing.'"

39   Howard Gardner, *The Mind's New Science: A History of the Cognitive Revolution* (New York: Basic Books, 1985), 144.

40   Ceruzzi, *Reckoners*, 94–97.

41   Aspray, *Computing Before Computers*, 204–206; K. Zuse et al., *The Computer: My Life* (Berlin, Heidelberg: Springer, 1993), 81.

42   J.A.N. Lee, "Howard Hathaway Aiken," Computer Pioneers (Institute of Electrical and Electronics Engineers, 1995), https://history.computer.org/pioneers/aiken.html; Uta C Merzbach, "Interview with Grace Murray Hopper," Computer Oral History Collection, 1969–1973, 1977 (Washington, DC: National Museum of American History, 1969), 7–8.

43   Howard Aiken, A. G. Oettinger, and T. C. Bartee, "Proposed Automatic Calculating Machine," *IEEE Spectrum* 1, no. 8 (1964): 63–64, https://doi.org/10.1109/MSPEC.1964.6500770; Aspray, *Computing Before Computers*, 213–214.

44   Aiken, Oettinger, and Bartee, "Proposed Automatic Calculating Machine."

45   Aiken, Oettinger, and Bartee, "Proposed Automatic Calculating Machine," 68.

46   L. Heide, *Punched-Card Systems and the Early Information Explosion, 1880–1945*, Studies in Industry and Society (Johns Hopkins University Press, 2009), 27–33.

47   Lee, "Howard Aiken."

48   Lee, "Howard Aiken."

49   "Feeds, Speeds and Specifications," IBM Archives (IBM), accessed June 7, 2021, https://www.ibm.com/ibm/history/exhibits/markI/markI_feeds2.html.

50   "ASCC General Description," IBM Archives (IBM, 1945), https://www.ibm.com/ibm/history/exhibits/markI/markI_description.html.

51   "Job 558, International Business Machines, 1943–1953," Norman Bel Geddes Database (Harry Ransom Center), accessed June 16, 2021, https://norman.hrc.utexas.edu/nbgpublic/details.cfm?id=476.

52   Merzbach, "Interview with Grace Hopper," 7–8.

53   Lee, "Howard Aiken."

54   Ceruzzi, *Reckoners*, 66.

55   J.A.N. Lee, "Grace Brewster Murray Hopper," Computer Pioneers (Institute of Electrical and Electronics Engineers, 1995), https://history.computer.org/pioneers/hopper.html.

56   Doron Swade, "Babbage, Charles (1791–1871), Mathematician and

Computer Pioneer," Oxford Dictionary of National Biography, 2009, https://doi.org/10.1093/ref:odnb/962.

57 Betty Alexandra Toole, "Byron, (Augusta) Ada [Married Name (Augusta) Ada King, Countess of Lovelace] (1815–1852), Mathematician and Computer Pioneer," *Oxford Dictionary of National Biography*, 2017, https://doi.org/10.1093/ref:odnb/37253.

58 Shigeru Takahashi, "Nakashima Akira," Japanese Computer Pioneers (IPSJ Computer Museum), accessed June 16, 2021, http://museum.ipsj.or.jp/en/pioneer/a-naka.html; Toma Kawanishi, "Prehistory of Switching Theory in Japan: Akira Nakashima and His Relay-Circuit Theory," *Historia Scientiarum* 29, no. 1 (2019): 136–162, https://jglobal.jst.go.jp/en/detail?JGLOBAL_ID=201902215823965846.

59 "ETL Mark I Relay-Based Automatic Computer," IPSJ Computer Museum (Information Processing Society of Japan), accessed June 21, 2021, http://museum.ipsj.or.jp/en/computer/dawn/0005.html; "ETL Mark II Relay-Based Automatic Computer," IPSJ Computer Museum (Information Processing Society of Japan), accessed June 21, 2021, http://museum.ipsj.or.jp/en/computer/dawn/0009.html.

60 "ETL Mark II"; Michael R. Williams, *A History of Computing Technology* (Los Alamitos, CA: IEEE Computer Society Press, 1997), 243–245.

61 "Casio 14-A," IPSJ Computer Museum (Information Processing Society of Japan), accessed June 21, 2021, http://museum.ipsj.or.jp/en/computer/dawn/0069.html.

62 Georges Bogaert, "Casio 14-A," Casio Calculator Collectors, accessed June 21, 2021, https://www.casio-calculator.com/Museum/Pages/_Numbers/14-A/Casio%2014-A.html.

63 Katsunori Kadokura, "The History of Japanese Calculators," X-Number, April 10, 2000, https://www.xnumber.com/xnumber/japanese_calculators.htm; "Establishment of CASIO."

64 J. P. Pederson, *International Directory of Company Histories*, Gale Virtual Reference Library (Detroit: St. James Press, 2001), 91; "CASIOの礎を築いた希代の発明家・樫尾俊雄【前編 (Toshio Kashio, a Rare Inventor Who Laid the Foundation of CASIO)," EMIRA (EMIRA Editorial Committee, June 21, 2017), https://emira-t.jp/ejinden/1775/; Raskin, "Casio's Eccentric Product Culture, Built on Embracing Failure."

65 "Casio 14-A."

66 "Casio 14-B," IPSJ Computer Museum (Information Processing Society of Japan), accessed June 21, 2021, http://museum.ipsj.or.jp/en/heritage/14 -B.html.

67 Raskin, "Casio's Eccentric Product Culture, Built on Embracing Failure"; "Relay Type Computer AL-1," IPSJ Computer Museum (Information Processing Society of Japan), accessed June 21, 2021, https://museum.ipsj.or .jp/en/heritage/Relay_type_computer_AL-1.html.

68 Ceruzzi, *Reckoners*, 97–98.

69 Aspray, *Computing Before Computers*, 219–220.

70 Ceruzzi, *Reckoners*, 97–98.

71 Aspray, *Computing Before Computers*, 220.

72 Merzbach, "Interview with Grace Hopper," 13.

73 "September 9: First Instance of Actual Computer Bug Being Found," This Day in History (Mountain View, CA: Computer History Museum), accessed June 21, 2021, https://www.computerhistory.org/tdih/september/9/.

74 "Establishment of CASIO."

8. THE SUMLOCK ANITA

1 Nigel Tout, ed., "Origins of the Bell Punch Company," Anita Calculators, accessed June 28, 2021, http://anita-calculators.info/html/origins_of_ bell_punch_co_.html.

2 "IBM Mainframes," IBM Archives (IBM), accessed June 28, 2021, https://www.ibm.com/ibm/history/exhibits/mainframe/mainframe_intro .html; Nigel Tout, "All Change at Bell Punch," Anita Calculators, accessed June 28, 2021, http://anita-calculators.info/html/all_change_at_bell_punch .html.

3 Nigel Tout, "The Development of ANITA: Part 1," Anita Calculators, accessed June 28, 2021, http://www.anita-calculators.info/html/development_ of_anita_1.html; "Nomination Form: Bell Telephone Laboratories," *National Register of History Places* (National Park Service, March 5, 1975).

4 Paul M. Connolly and Robert W. Eisenmenger, "The Role of the Federal Reserve in the Payments System," *Federal Reserve Bank of Boston Conference Series* 45 (October 2000): 131–161, https://ideas.repec.org/a/fip/fedbcp/ y2000ioctp131-161n45.html.

5 Tout, "Origins of the Bell Punch Company."

6   Nigel Tout, "Introduction to Plus & Sumlock Mechanical Calculators," Anita Calculators, accessed June 29, 2021, http://anita-calculators.info/html/introduction1.html.

7   Roger G. Johnson, "Andrew D. Booth: Britain's Other 'Fourth Man,'" in *IFIP Advances in Information and Communication Technology*, vol. 325 (New York: Springer, 2010), 24, https://doi.org/10.1007/978-3-642-15199-6_4; Simon H. Lavington, *Early British Computers : The Story of Vintage Computers and the People Who Built Them* (Manchester University Press, 1980), chap. 5.

8   B. E. Carpenter and R. W. Doran, *A. M. Turing's ACE Report of 1946 and Other Papers* (Cambridge, MA: Massachusetts Institute of Technology, 1986), 126.

9   Nick Pelling, "The Case for the First Business Computer" (Kingston University Business School, March 26, 2002), https://www.nickpelling.com/Leo1.html; Lavington, *Early British Computers*, chap. 13.

10   Nigel Tout, "ANITA: The World's First Electronic Desktop Calculator," Anita Calculators, accessed June 29, 2021, http://anita-calculators.info/html/summary.html.

11   Johnson, "Andew D. Booth," 24, 30.

12   Reese V. Jenkins, "Edison, Thomas Alva (1847–1931), Inventor and Business Entrepreneur," American National Biography, February 2000, https://doi.org/10.1093/anb/9780198606697.article.1300470; "Samuel and Nancy Elliott Edison," Thomas Edison National Historical Park (U.S. National Park Service, February 26, 2015), https://www.nps.gov/edis/learn/historyculture/samuel-and-nancy-elliott-edison.htm.

13   Jenkins, "Edison, Thomas Alva"; Burns, "Bell, Alexander Graham."

14   Greg Sanford, "Illuminating Systems: Edison and Electrical Incandescence," *OAH Magazine of History* 4, no. 2 (1989): 18–19; Espen Skjong et al., "The Marine Vessel's Electrical Power System: From Its Birth to Present Day," *Proceedings of the IEEE* 103 (2015): 2–3, https://doi.org/10.1109/JPROC.2015.2496722; "Electric Light and Power System," Thomas A. Edison Papers (Rutgers School of Arts and Sciences, October 28, 2016), http://edison.rutgers.edu/power.htm.

15   "Edison's Patents," Thomas A. Edison Papers (Rutgers School of Arts and Sciences), accessed June 23, 2021, http://edison.rutgers.edu/patents.htm.

16   Clayton H. Sharp, "The Edison Effect and Its Modern Applications,"

Journal of the American Institute of Electrical Engineers 41, no. 1 (December 12, 1922): 68–69, https://doi.org/10.1109/joaiee.1922.6594383.

17   Sharp, "Edison Effect, " 69.

18   Isobel Falconer, "Corpuscles, Electrons and Cathode Rays: J. J. Thomson and the 'Discovery of the Electron,'" *The British Journal for the History of Science* 20, no. 3 (1987): 241–276.

19   Thomas A Edison, Electrical Indicator, 307,031 A (USA, issued October 1884), http://www.google.com/patents/US307031A.

20   Sharp, "Edison Effect," 69.

21   John Ambrose Fleming, Instrument for converting alternating electric currents into continuous currents, 803,684 A (USA, issued April 1905), http://www.google.com/patents/US803684A; J. T. MacGregor-Morris and Graeme J. N. Gooday, "Fleming, Sir (John) Ambrose (1849–1945), Electrical Engineer and University Teacher," *Oxford Dictionary of National Biography*, 2011, https://doi.org/10.1093/ref:odnb/33170; H. F. Dylla and Steven T. Corneliussen, "John Ambrose Fleming and the Beginning of Electronics," *Journal of Vacuum Science & Technology A: Vacuum, Surfaces, and Films* 23, no. 4 (July 2005): 8, https://doi.org/10.1116/1.1881652.

22   Gerald F. Tyne, Diana D. Menkes, and Elliot N. Sivowitch, *Saga of the Vacuum Tube* (Indianapolis: Prompt, 1994), 75–76.

23   Christopher H. Sterling, "De Forest, Lee (1873–1961), Radio Engineer and Inventor," American National Biography, 2000, https://doi.org/10.1093/anb/9780198606697.article.1300416.

24   Tyne, Menkes, and Sivowitch, *Saga of the Vacuum Tube*, 75–76.

25   Karl R. Spangenberg, *Vacuum Tubes* (New York: McGraw Hill, 1948), 6–7.

26   Spangenberg, *Vacuum Tubes*, 2–3.

27   Rick Bensene, "Friden EC-130 Electronic Calculator," Old Calculator Museum, August 9, 2020, https://www.oldcalculatormuseum.com/friden130.html.

28   Bensene, "Friden EC-130."

29   Robert B Tomer, *Getting the Most out of Vacuum Tubes* (Indianapolis: H. W. Sams, 1960), chap. 1.

30   Ceruzzi, *Reckoners*, 97–98.

31   Aspray, *Computing Before Computers*, 220.

32   W. H. Eccles and F. W. Jordan, "A Trigger Relay Utilizing Three-Electrode Thermionic Vacuum Tubes," *The Electrician* 83 (1919): 298.

33   Aspray, *Computing Before Computers*, 225–226.

34   John L. Gustafson, "Atanasoff, John Vincent (1903–1995), Inventor and Computer Scientist," American National Biography, 2002, https://doi.org/10.1093/anb/9780198606697.article.1302645; Allan R. Mackintosh, "Dr. Atanasoff's Computer," *Scientific American* 259, no. 2 (1988): 90–96.

35   J.A.N. Lee, "John Vincent Atanasoff," Computer Pioneers (Institute of Electrical and Electronics Engineers, 2012), https://history.computer.org/pioneers/atanasoff.html.

36   Lee, "John Vincent Atanasoff"; Aspray, *Computing Before Computers*, chap. 5.

37   Lee, "John Vincent Atanasoff."

38   Lee, "John Vincent Atanasoff."

39   Ceruzzi, *Reckoners*, 108–10.

40   J. J. O'Connor and E. F. Robertson, "Lewis Fry Richardson," MacTutor (School of Mathematics and Statistics, University of St. Andrews, 2003), https://mathshistory.st-andrews.ac.uk/Biographies/Richardson/; Angelo Vulpiani, "Lewis Fry Richardson: Scientist, Visionary and Pacifist," *Lettera Matematica* 2, no. 3 (September 9, 2014): 121–128, https://doi.org/10.1007/s40329-014-0063-z.

41   Ceruzzi, *Reckoners*, 109–124.

42   Kennedy Jr., "Electronic Computer Flashes Answers."

43   "Colossus" (The National Museum of Computing), accessed June 30, 2021, https://www.tnmoc.org/colossus; Michael Williams, "The First Public Discussion of the Secret Colossus Project," *IEEE Annals of the History of Computing* 40, no. 1 (January 1, 2018): 84–87, https://doi.org/10.1109/MAHC.2018.012171272.

44   Carpenter and Doran, *ACE Report and Other Papers*, 5–6; J. A. N. Lee, "Andrew Donald Booth," Computer Pioneers (Institute of Electrical and Electronics Engineers, 2012), https://history.computer.org/pioneers/booth-ad.html.

45   Alan Sobel, "Electronic Numbers," *Scientific American* 228, no. 6 (1973): 64–73; Nigel Tout, "ANITA Mk 8 Calculator," Anita Calculators, accessed July 1, 2021, http://www.anita-calculators.info/html/anita_mk_8.html.

46   Tout "ANITA Mk 8 Calculator"; Tout, "Development of ANITA: Part 1."

47   Tout, "ANITA Mk 8 Calculator."

48   "Review of the ANITA Mk 8," Anita Calculators, accessed June 29, 2021, http://anita-calculators.info/html/review_of_anita_mk_8.html.

49   proxxima038, "Anita MK8 Vintage Electronic Desktop Calculator Demonstration," YouTube, August 7, 2014, https://www.youtube.com/watch?v=6-qvtiqWens.

50   Nigel Tout, "ANITA Is Launched," Anita Calculators, accessed July 1, 2021, http://www.anita-calculators.info/html/anita_is_launched.html; "Review of the ANITA Mk 8."

51   John Sparkes, "The Development of the ANITA Calculators at the Bell Punch Company," 2001, 1, http://anita-calculators.info/JohnSparkes -AnitaCalculators.pdf.

52   John F. Chown, *A History of Money: From AD 800* (London: Routledge, 1994), 17–18.

53   Chown, *History of Money*, 65–66.

54   N.E.A. Moore, *The Decimalisation of Britain's Currency* (London: H.M. Stationery Office, 1973), 2–5.

55   Nigel Tout, "Sterling Currency Calculators," Vintage Calculators Web Museum, accessed July 2, 2021, http://www.vintagecalculators.com/html/ sterling_calculators.html.

56   Tony Beaumont, "Price War in the Calculator Business," *New Scientist*, 1972, 748; Nigel Tout, "The Development of ANITA: Part 2," Anita Calculators, accessed July 2, 2021, http://www.anita-calculators.info/html/ development_of_anita_2.html.

57   Beaumont, "Price War in the Calculator Business," 748–749.

58   "ANITA Developed Further," Anita Calculators, n.d., http://anita -calculators.info/html/anita_developed_further.html.

59   Personal correspondence with Nigel Tout by the author, 2021.

60   Rick Bensene, "Sumlock Comptometer/Bell Punch Anita C/ VIII," The Old Calculator Museum, December 11, 2019, https://www .oldcalculatormuseum.com/anitaC-VIII.html.

61   N. Kitz, "Cold Cathode Trigger and Counter Tubes for Computing Applications," in *Cold Cathode Tubes and Their Applications* (University of Cambridge, 1964), III/7/12.

9. THE OLIVETTI PROGRAMMA 101

1   Oscar Ornati, "The Italian Economic Miracle and Organized Labor," *Social Research* 30, no. 4 (1963): 519–526.

2   Meryle Secrest, *The Mysterious Affair at Olivetti: IBM, the CIA, and the*

Cold War Conspiracy to Shut down Production of the World's First Desktop Computer (New York: Alfred A. Knopf, 2019), 1, 11.

3   Alfred R. Zipser, "Olivetti Hopes to Rejuvenate Underwood, Typewriter Pioneer; Gains Foreseen for Underwood," *New York Times*, June 12, 1960, https://www.nytimes.com/1960/06/12/archives/olivetti-hopes-to-rejuvenate-underwood-typewriter-pioneer-gains-for.html; Secrest, *The Mysterious Affair at Olivetti*, 11–12.

4   "Olivetti," Jewish Virtual Library (American–Israeli Cooperative Enterprise), accessed July 22, 2021, https://www.jewishvirtuallibrary.org/olivetti; Secrest, *Mysterious Affair at Olivetti*, 49–52.

5   "Adriano Olivetti, 1901–1960" (Fondazione Adriano Olivetti, 1962), https://www.fondazioneadrianolivetti.it/bioadrianoolivetti/.

6   Secrest, *Mysterious Affair at Olivetti*, 49–52.

7   Secrest, *Mysterious Affair at Olivetti*, 58–59, 68.

8   Secrest, *Mysterious Affair at Olivetti*, 87.

9   Allen Dulles et al., *From Hitler's Doorstep: The Wartime Intelligence Reports of Allen Dulles, 1942–1945* (University Park: Pennsylvania State University Press, 1996), 72–73.

10   Secrest, *Mysterious Affair at Olivetti*, 111–121.

11   Emilio Renzi, "L'avventura Olivetti," Enciclopedia Italiana (Rome: Istituto della Enciclopedia Italiana, 2013), https://www.treccani.it/enciclopedia/l-avventura-olivetti_%28Il-Contributo-italiano-alla-storia-del-Pensiero:-Tecnica%29/.

12   Secrest, *Mysterious Affair at Olivetti*, 138–139.

13   S. Lucamante, *Italy and the Bourgeoisie: The Re-Thinking of a Class* (Vancouver, BC: Fairleigh Dickinson University Press, 2009), 90–91.

14   Secrest, *Mysterious Affair at Olivetti*, 37–39.

15   "Adriano Olivetti, 1901–1960"; Nikil Saval, "Utopia, Abandoned," *New York Times*, September 19, 2019, https://www.nytimes.com/2019/08/28/t-magazine/olivetti-typewriters-ivrea-italy.html; Annalisa Merelli, "This Italian Company Pioneered Innovative Startup Culture—in the 1930s," Quartz, August 11, 2015, https://qz.com/455328/this-italian-company-pioneered-innovative-startup-company-in-the-1930s/.

16   "Ivrea, Industrial City of the 20th Century," World Heritage List (UNESCO, 2018), https://whc.unesco.org/en/list/1538/.

17   Saval, "Utopia, Abandoned."

18   Secrest, *Mysterious Affair at Olivetti*, 28–29.

19   "Adriano Olivetti, 1901–1960."

20   "Poster, Olivetti, 1976 (1979-42-5)," Collection of Cooper Hewitt, Smithsonian Design Museum (New York: Smithsonian Institution), accessed July 23, 2021, https://collection.cooperhewitt.org/objects/18498105/.

21   Secrest, *Mysterious Affair at Olivetti*, 85–86; Arthur Molella, "The Italian Soul of Steve Jobs," National Museum of American History (Washington, DC: Smithsonian Institution, January 24, 2012), https://americanhistory.si .edu/blog/2012/01/the-italian-soul-of-steve-jobs.html.

22   Leo Lionni, *Olivetti, Design in Industry* (New York: Museum of Modern Art, 1952), 3, www.moma.org/calendar/exhibitions/2741.

23   Saval, "Utopia, Abandoned."

24   Richard Polt, "Writers and Their Typewriters," The Classic Typewriter Page, accessed July 23, 2021, https://site.xavier.edu/polt/typewriters/typers .html.

25   Campbell Bickerstaff, "Olivetti Divisumma 24 Electric Calculator (K456)," Museum of Applied Arts & Sciences, 2013, https://collection.maas .museum/object/262195.

26   Federico Barbiellini Amidei, Andrea Goldstein, and Marcella Spadoni, "European Acquisitions in the United States: Re-Examining Olivetti-Underwood Fifty Years Later," IDEAS, March 2010, 12, https://ideas.repec .org/p/bdi/workqs/qse_2.html.

27   Secrest, *Mysterious Affair at Olivetti*, 201–202.

28   Zipser, "Olivetti Hopes to Rejuvenate Underwood"; Renzi, "L'avventura Olivetti."

29   "Adriano Olivetti, 1901–1960"; Secrest, *The Mysterious Affair at Olivetti*, 13–14.

30   "Corporations: Olivetti Moves In," *Time*, April 25, 1960, http://content .time.com/time/subscriber/article/0,33009,826332,00.html.

31   "Milestones," *Time*, March 7, 1960, http://content.time.com/time/ subscriber/article/0,33009,894747,00.html.

32   Giuseppe Rao, "La Sfida Al Futuro Di Adriano e Roberto Olivetti. Il Laboratorio Di Ricerche Elettroniche, Mario Tchou e l'Elea 9003.," *Mélanges de l'École Française de Rome* 115, no. 2 (2003): 646–650, 655, https://doi .org/10.3406/mefr.2003.10059.

33   Elisabetta Mori, "The Italian Computer: Italy's Olivetti Was an Early

Pioneer of Digital Computers and Transistors," *IEEE Spectrum* 56, no. 6 (June 1, 2019): 40–47, https://doi.org/10.1109/MSPEC.2019.8727145.

34 "Personnel: Changes of the Week," *Time*, March 31, 1960, http://content .time.com/time/subscriber/article/0,33009,894839,00.html; Renzi, "L'avventura Olivetti."

35 Amidei, Goldstein, and Spadoni, "Re-Examining Olivetti-Underwood Fifty Years Later," 17–18.

36 "Olivetti Summa Prima 20" (Cambridge: Centre for Computing History), accessed July 29, 2021, http://www.computinghistory.org.uk/det/6012/ Olivetti-Summa-Prima-20/.

37 *Spot - Olivetti - Calcolatore - Programma 101* (Archivio Nazionale Cinema d'Impresa, 1966), https://www.youtube.com/watch?v=WnItIQSwfSw; Secrest, *Mysterious Affair at Olivetti*, 211.

38 Riccardo Bianichini, "Olivetti Programma 101: At the Origins of the Personal Computer," Inexhibit (Bianchini & Lusiardi Associati, November 1, 2019), https://www.inexhibit.com/case-studies/olivetti-programma-101-at-the -origins-of-the-personal-computer/.

39 Rao, "La Sfida Al Futuro Di Adriano e Roberto Olivetti," 676.

40 Giuditta Parolini, "Olivetti Elea 9003: Between Scientific Research and Computer Business," in *History of Computing and Education 3 (HCE3)*, ed. John Impagliazzo (New York: Springer, 2008), 40.

41 Secrest, *Mysterious Affair at Olivetti*, 193–194; Luigi Logrippo, "My First Two Computers: Elea 9003 and Elea 6001," April 2019, https://www.site .uottawa.ca/~luigi/papers/elea.htm.

42 Julius Edgar Lilienfeld, Method and apparatus for controlling electric currents, issued October 1926, http://www.google.com/patents/US1745175A.

43 Lilienfeld; "1926: Field Effect Semiconductor Device Concepts Patented," The Silicon Engine (Mountain View, CA: Computer History Museum), accessed July 29, 2021, https://www.computerhistory.org/siliconengine/field -effect-semiconductor-device-concepts-patented/.

44 Michael Riordan, "The Lost History of the Transistor," *IEEE Spectrum* 41, no. 5 (2004): 44–49, https://doi.org/10.1109/MSPEC.2004.1296014.

45 W. F. Brinkman, D. E. Haggan, and W. W. Troutman, "A History of the Invention of the Transistor and Where It Will Lead Us," *IEEE Journal of Solid-State Circuits* 32, no. 12 (1997): 1858–1861, https://doi.org/10.1109/4 .643644.

46 "The Nobel Prize in Physics 1956" (Nobel Prize Outreach AB, 2021), https://www.nobelprize.org/prizes/physics/1956/summary/.

47 Brinkman, Haggan, and Troutman, "History of the Invention of the Transistor," 1860; Tom Wolfe, "The Tinkerings of Robert Noyce: How the Sun Rose on the Silicon Valley," Esquire, 1983, https://www.esquire.com/news -politics/a12149389/robert-noyce-tom-wolfe/.

48 Ceruzzi, Reckoners, 99.

49 Thomas A. Fackler et al., "How Antitrust Enforcement Can Spur Inno-vation: Bell Labs and the 1956 Consent Decree" (London, January 2017), 1–2, https://cepr.org/active/publications/discussion_papers/dp.php?dpno=11793; Kevin Granville and Tiffany Hsu, "AT&T Has Had Many Run-Ins with the Government," New York Times, June 12, 2018, https://www.nytimes .com/2018/06/12/business/dealbook/att-antitrust.html.

50 "Trends in the Semiconductor Industry," Semiconductor History Museum of Japan (Tokyo: Society of Semiconductor Industry Specialists), accessed July 29, 2021, https://www.shmj.or.jp/english/trends.html; Mori, "The Italian Computer."

51 Campbell Bickerstaff, "Olivetti Programma 101 Computer Designed by Mario Bellini (2008/107/1)," Museum of Applied Arts & Sciences, 2008, https://collection.maas.museum/object/378406; "Olivetti Programma 101," Curtamania, July 1, 2020, http://www.curtamania.com/curta/database/ brand/olivetti/Olivetti Programma 101/index.html.

52 Francesco Bonomi, "Olivetti Programma 101," 1999, The electronics, http://www.silab.it/frox/p101/_index.html.

53 Bonomi, "Olivetti Programma 101," The magnetic card.

54 Bonomi, "Olivetti Programma 101," The P101 Instructions.

55 Bonomi, "Olivetti Programma 101," Logical instructions.

56 Pier Giorgio Perotto and Giovanni de Sandre, Program controlled electronic computer, 3,495,222 A (USA, issued October 1970), http://www .google.com/patents/US3495222A.

57 Nigel Tout, "Olivetti-Underwood Programma 101," Vintage Calculators Web Museum, accessed July 30, 2021, http://www.vintagecalculators.com/ html/olivetti_programma_101.html.

58 Bonomi, "Olivetti Programma 101," The Programma 101 Memory module.

59 Cara McCarty, Mario Bellini, Designer (New York: Museum of Modern Art, 1987), 10, https://www.moma.org/calendar/exhibitions/1785.

60   McCarty, *Bellini*, 8.

61   "Calcolatore Progammabile Da Tavolo - Olivetti Programma 101 (15801)" (Milan: Museo Nazionale della Scienza e della Tecnologia Leonardo da Vinci), accessed August 5, 2021, http://www.museoscienza.it/dipartimenti/catalogo_collezioni/scheda_oggetto.asp?idk_in=ST010-00250&arg=programma%101.

62   Rick Bensene, "Olivetti Programma 101 Electronic Calculator," Old Calculator Museum, September 22, 2019, https://www.oldcalculatormuseum.com/friden130.html.

63   McCarty, *Bellini*, 14–15.

64   "Divisumma 18 Calculator, 1973 (1986-99-41)," Collection of Cooper Hewitt, Smithsonian Design Museum (New York: Smithsonian Institution), accessed August 5, 2021, https://collection.cooperhewitt.org/objects/18621873/; "Divisumma 18 Electronic Printing Calculator (450.1974.1-2)" (New York: Museum of Modern Art, 1972), https://www.moma.org/collection/works/3805.

65   Rick Bensene, "Mathatronics Mathatron 8-48M Mod II Electronic Calculator," Old Calculator Museum, February 26, 2017, https://www.oldcalculatormuseum.com/c-math8-48m.html.

66   Nigel Tout, "IME 84rc," Vintage Calculators Web Museum, accessed August 5, 2021, http://www.vintagecalculators.com/html/ime_84rc.html.

67   "G.E. Completes Olivetti Deal for Electronic-Processing Unit," *New York Times*, September 1, 1964, https://www.nytimes.com/1964/09/01/archives/ge-completes-olivetti-deal-for-electronicprocessing-unit.html.

68   Secrest, *Mysterious Affair at Olivetti*, 215–216.

69   "Alle Origini Del Personal Computer: L'Olivetti Programma 101," Personal Computing (Ivrea: Associazione Archivio Storico Olivetti), accessed July 30, 2021, https://www.storiaolivetti.it/articolo/87-alle-origini-del-personal-computer-lolivetti-pr/.

70   Secrest, *Mysterious Affair at Olivetti*, 215–216.

71   Bensene, "Olivetti Programma 101 Electronic Calculator"; Bianichini, "Olivetti Programma 101."

72   "9100A Desktop Calculator, 1968," Innovation Gallery (Hewlett-Packard Enterprise), accessed July 30, 2021, https://www.hpe.com/us/en/about/history/innovation-gallery/021-product.html.

73   Secrest, *Mysterious Affair at Olivetti*, 221–222.

74   "CPI Inflation Calculator"; Tout, "Olivetti-Underwood Programma

101"; Douglas W. Jones, "Frequently Asked Questions," The Digital Equipment Corporation PDP-8, accessed August 5, 2021, http://homepage.cs.uiowa.edu/~jones/pdp8/faqs/; Douglas W. Jones, "Index," The Digital Equipment Corporation PDP-8, accessed August 5, 2021, http://homepage.cs.uiowa.edu/~jones/pdp8/.

75 Docabout and Zenit Artivisive, *Programma 101—Memory of the Future* (101 Project, 2010), https://www.youtube.com/watch?v=lpkqdbz1R_s.

76 Amidei, Goldstein, and Spadoni, "Re-Examining Olivetti-Underwood Fifty Years Later," 27.

77 Amidei, Goldstein, and Spadoni, "Re-Examining Olivetti-Underwood Fifty Years Later," 28.

78 *Spot - Olivetti - Calcolatore - Programma 101.*

79 Sandra Johnson, "Interview with David W. Whittle," *NASA Johnson Space Center Oral History Project* (Houston, TX: National Aeronautical and Space Administration, February 16, 2006), https://historycollection.jsc.nasa.gov/JSCHistoryPortal/history/oral_histories/WhittleDW/WhittleDW_2-16-06.htm.

80 William Shawcross, "The Secret Bombing of Cambodia," in *Light at the End of the Tunnel: A Vietnam War Anthology*, ed. Andrew Jon Rotter (New York: St. Martin's Press, 1991), 206–215.

81 Amidei, Goldstein, and Spadoni, "Re-Examining Olivetti-Underwood Fifty Years Later," 25–33.

82 Amidei, Goldstein, and Spadoni, "Re-Examining Olivetti-Underwood Fifty Years Later," 28.

83 Bickerstaff, "Olivetti Programma 101."

84 Corrado Bonfanti, "Information Technology in Italy: The Origins and the Early Years (1954–1965)," *IFIP Advances in Information and Communication Technology* AICT-387 (2012): 1–3, https://doi.org/10.1007/978-3-642-33899-1_19; Mori, "The Italian Computer."

85 Secrest, *Mysterious Affair at Olivetti*, 246.

86 Secrest, *Mysterious Affair at Olivetti*, 207–208.

87 Secrest, *Mysterious Affair at Olivetti*, 218–220.

10. THE TEXAS INSTRUMENTS CAL TECH

1 Jack S. Kilby, "Invention of the Integrated Circuit," *IEEE Transactions on Electron Devices* 23, no. 7 (1976): 648, https://doi.org/10.1109/T-ED.1976

.18467; Michael Wolff, "Interview with Jack St. Clair Kilby" (IEEE History Center, December 2, 1975), sec. Miniaturization and Molecular Electronics, https://ethw.org/Oral-History:Jack_Kilby.

2   Kilby, "Invention of the Integrated Circuit," 650.

3   Kilby, 648; "B-29 Superfortress" (Boeing), accessed August 25, 2021, https://www.boeing.com/history/products/b-29-superfortress.page.

4   A. Belous and V. Saladukha, *High-Speed Digital System Design: Art, Science and Experience* (Cham: Springer, 2019), 544–546; "Integrated Circuits Come of Age" (Washington, DC: US Air Force Systems Command, 1966), 3, https://www.computerhistory.org/collections/catalog/102740473.

5   Albert Parker Hanson, Improvements in or connected with Electric Cables and the Jointing of the same, GB190304681 (A) (UK, issued 1904).

6   Belous and Saladukha, *High-Speed Digital System Design*.

7   T. R. Reid, "The Texas Edison," *Texas Monthly*, July 1982, sec. The Death of the Tube, https://www.texasmonthly.com/articles/the-texas-edison/.

8   Gary Price, "Texas Instruments," Handbook of Texas Online (Texas State Historical Association), accessed August 20, 2021, https://www.tshaonline.org/handbook/entries/texas-instruments; Reid, "The Texas Edison," sec. Chips and Money.

9   Reid, sec. Chips and Money.

10   Michael Riordan, "The Lost History of the Transistor," *IEEE Spectrum* 41, no. 5 (2004): 44–49, https://doi.org/10.1109/MSPEC.2004.1296014.

11   Andrew Goldstein, "Interview with Gordon K. Teal" (IEEE History Center, December 1991), sec. Speech: American Academy of Achievement, Texas Instruments, https://ethw.org/Oral-History:Gordon_K._Teal.

12   Riordan, "The Lost History of the Transistor."

13   Goldstein, "Interview with Gordon K. Teal," sec. Texas Instruments; Riordan, "The Lost History of the Transistor."

14   Reid, "The Texas Edison."

15   Wolff, "Interview with Jack St. Clair Kilby," sec. Conception of Integrated Circuit.

16   Wolff, sec. Development of Prototypes; Kilby, "Invention of the Integrated Circuit," 650.

17   Kilby, "Invention of the Integrated Circuit," 652.

18   Wolfe, "Tinkerings of Robert Noyce."

19   "Integrated Circuits Come of Age," 4.

20   Jack S. Kilby, "Jack Kilby Papers, 1923–2005: A Guide to the Collection," accessed August 27, 2021, https://legacy.lib.utexas.edu/taro/smu/00116/smu-00116.html; "Integrated Circuits Come of Age," 10.

21   "A Molecular Electronic Computer by TI" (Texas Instruments, 1961).

22   G.W.A. Dummer and J. Mackenzie Robertson, eds., "Texas Instruments Series 51 Semiconductor Networks," in *American Microelectronics Data Annual 1964–65* (Pergamon, 1964), 596–797, https://doi.org/10.1016/B978-0-08-011064-6.50041-X.

23   "Developing the First ICs to Orbit Earth" (Texas Instruments, June 24, 2021), https://news.ti.com/blog/2021/06/24/developing-first-ics-to-orbit-earth.t

24   Sue Moore, "News Release" (Dallas, TX: Texas Instruments, March 10, 1961), https://digitalcollections.smu.edu/digital/collection/tir/id/291/rec/100; Don MacIver, "News Release" (Dallas, TX: Texas Instruments, January 1962), https://digitalcollections.smu.edu/digital/collection/tir/id/296; R. G. Davies, "News Release" (Dallas, TX: Texas Instruments, September 12, 1962), https://digitalcollections.smu.edu/digital/collection/tir/id/302; Clark W. Fischel, "News Release" (Dallas, TX: Texas Instruments, June 12, 1962), https://digitalcollections.smu.edu/digital/collection/tir/id/3305.

25   "Integrated Circuits Come of Age," 21.

26   Kathy B. Hamrick, "The History of the Hand-Held Electronic Calculator," *American Mathematical Monthly*, no. 102 (1996): 633.

27   Paul D. Davis and Diana Kleiner, "The Breakthrough Breadboard Feasibility Model: The Development of the First All-Transistor Radio," *The Southwestern Historical Quarterly* 97, no. 1 (1993): 57; Hamrick, "The Hand-Held Electronic Calculator," 634.

28   Davis and Kleiner, "Breakthrough Breadboard Feasibility Model," 57–61.

29   Hamrick, "The Hand-Held Electronic Calculator," 634; Davis and Kleiner, "Breakthrough Breadboard Feasibility Model," 57–61; "Income of Persons in the United States: 1954" (United States Census Bureau, October 1955), https://www.census.gov/library/publications/1955/demo/p60-019.html.

30   Hamrick, "The Hand-Held Electronic Calculator," 635.

31   "Income of Persons in the United States: 1954" (United States Census Bureau, October 1955), https://www.census.gov/library/publications/1955/demo/p60-019.html.

32 Tout, "IME 84rc."

33 Thomas M. Okon, "The First Handheld Digital Calculator Celebrates 50 Years, Part 1," Electronic Design, March 27, 2017, https://www.electronicdesign.com/technologies/analog/article/21804824/the-first-handheld-digital-calculator-celebrates-50-years-part-1.

34 Cade Metz, "Jerry Merryman, Co-Inventor of the Pocket Calculator, Dies at 86," New York Times, March 7, 2017, https://www.nytimes.com/2019/03/07/obituaries/jerry-merryman-dead.html.

35 Okon, "First Handheld Digital Calculator, Part 1."

36 Metz, "Jerry Merryman."

37 Rick Bensene, "Canon Pocketronic Electronic Calculator," The Old Calculator Web Museum, January 9, 2011, https://www.oldcalculatormuseum.com/pocketronic.html.

38 Bensene, "Canon Pocketronic."

39 Okon, "First Handheld Digital Calculator, Part 1."

40 Hamrick, "Hand-Held Electronic Calculator," 634.

41 Bensene, "Canon Pocketronic."

42 "Adding Machines - Ten Keys & Fewer."

43 R. L. Deininger, "Human Factors Engineering Studies of the Design and Use of Pushbutton Telephone Sets," Bell System Technical Journal 39, no. 4 (July 29, 1960): 995–1012, https://doi.org/10.1002/j.1538-7305.1960.tb04447.x.

44 Hamrick, "Hand-Held Electronic Calculator," 634; Thomas M. Okon and James R. Biard, "The First Practical LED," Edison Tech Center 9 (2015).

45 Bensene, "Canon Pocketronic"; Thomas M. Okon, "The First Handheld Digital Calculator Celebrates 50 Years, Part 2," Electronic Design, April 5, 2017, https://www.electronicdesign.com/technologies/analog/article/21804872/the-first-handheld-digital-calculator-celebrates-50-years-part-2.

46 Bensene, "Canon Pocketronic."

47 Hamrick, "Hand-Held Electronic Calculator," 635.

48 Hamrick, "Hand-Held Electronic Calculator," 635.

49 Joerg Woerner, personal correspondence with the author, September 24, 2021.

50 Jerry Hirsch, "First Volkswagen Beetle Arrived in a U.S. Showroom 65 Years Ago," Los Angeles Times, January 31, 2014, https://www.latimes.com/

business/la-xpm-2014-jan-31-la-fi-hy-volkswagen-beetle-65-years-20140130
-story.html.

51 Ira A. Robbins, "British Invasion," in *Encyclopaedia Britannica*, 2020,
https://www.britannica.com/event/British-Invasion.

52 "Japanese Industrial Policy: The Postwar Record and the Case of Supercomputers," *Competing Economies: America, Europe, and the Pacific Rim*
(Washington, DC: U.S. Government Printing Office, 1991), 241.

53 Dylan Gerstel and Matthew P. Goodman, "Japan: Industrial Policy
and the Economic Miracle," *From Industrial Policy to Innovation Strategy: Lessons from Japan, Europe, and the United States*, Lessons from Japan,
Europe, and the United States (Center for Strategic and International Studies (CSIS), September 16, 2020).

54 William M. Tsutsui, "W. Edwards Deming and the Origins of Quality
Control in Japan," *Journal of Japanese Studies* 22, no. 2 (1996): 295–325.

55 Toyoguchi Katsuhei and Penny Bailey, "'Good Design' and 'Good
Quality' for the Consumer (1965)," *Review of Japanese Culture and Society* 28
(September 16, 2016): 144–151.

56 "Trends in the Semiconductor Industry."

57 Nigel Tout, "The Pocket Calculator Race," Vintage Calculators Web
Museum, accessed September 16, 2021, http://www.vintagecalculators.com/
html/the_pocket_calculator_race.html.

58 Dennis J. Encarnation and Mark Mason, "Neither MITI Nor America: The Political Economy of Capital Liberalization in Japan," *International Organization* 44, no. 1 (1990): 39; Thomas Hoeren, Francesca
Guadagno, and Sacha Wunsch-Vincent, *Breakthrough Technologies – Semiconductor, Innovation and Intellectual Property*, vol. 27 (Geneva: WIPO,
2015), 20.

59 Encarnation and Mason, "Neither MITI Nor America," 40.

60 Rick Bensene, "Sony SOBAX ICC-500W Electronic Calculator," The
Old Calculator Web Museum, accessed September 16, 2021, https://www
.oldcalculatormuseum.com/sony-sobax.html.

61 Rick Bensene, "Specifications for Sony Sobax ICC-400W," The Old
Calculator Web Museum, accessed September 16, 2021, https://www
.oldcalculatormuseum.com/s-sony400.html.

62 Nigel Tout, "Sharp QT-8D 'Micro Compet' and Clones," Vin-

tage Calculators Web Museum, accessed September 16, 2021, http://www
.vintagecalculators.com/html/sharp_qt-8d.html.

63 Encarnation and Mason, "Neither MITI Nor America," 44–45.

64 Yongdo Kim, "Interfirm Cooperation in Japan's Integrated Circuit
Industry, 1960s–1970s," *The Business History Review* 86, no. 4 (2012): 778.

65 Woerner, personal correspondence with the author.

66 Okon, "First Handheld Digital Calculator, Part 2."

67 "The History of Canon 1933–1961," Canon Global, accessed September 17, 2021, https://global.canon/en/corporate/history/01.html; "Film Cameras - 1960–1965," Canon Camera Museum, accessed September 17, 2021,
https://global.canon/en/c-museum/series_search.html?t=camera&s=film&y1=
1960&y2=1965.

68 Rick Bensene, "Canon 161 Calculator," The Old Calculator Web
Museum, accessed September 17, 2021, https://www.oldcalculatormuseum
.com/canon161.html.

69 Stephen W. Fields et al., "U.S. Firms Gird for Calculator Battle," *Electronics* (New York: McGraw Hill, November 1970), 85.

70 Okon, "First Handheld Digital Calculator, Part 2"; Bensene, "Canon
Pocketronic."

71 Hamrick, "Hand-Held Electronic Calculator," 635; Rick Bensene,
"Specifications for Canon Pocketronic," The Old Calculator Web Museum,
accessed September 17, 2021, https://www.oldcalculatormuseum.com/s
-pocketronic.html.

72 "Electronics Abroad," *Electronics* (New York: McGraw Hill, June 10,
1968), 257.

73 Arthur Fisher, "Carry This Calculator in Your Pocket," *Popular Science*
(New York: Popular Science, August 1970).

74 Nigel Tout, "Texas Instruments Cal-Tech," Vintage Calculators Web
Museum, accessed September 17, 2021, http://www.vintagecalculators.com/
html/ti_cal-tech1.html.

75 John R. Free, "Microelectronics Shrinks the Calculator," *Popular Science*
(New York: Popular Science, June 1971).

76 Grace Lichtenstein, "New Calculators Catch Public's Fancy," *New York
Times*, October 1972, https://www.nytimes.com/1972/10/28/archives/new
-calculators-catch-publics-fancy.html.

11. THE BUSICOM 141-PF

1   Leslie R. Berlin, "Robert Noyce and Fairchild Semiconductor, 1957–1968," *The Business History Review* 75, no. 1 (2001): 68–69; Gordon E. Moore, "Intel: Memories and the Microprocessor," *Daedalus* 125, no. 2 (1996): 57–59.

2   Annalee Saxenian, "The Genesis of Silicon Valley," *Built Environment (1978–)* 9, no. 1 (September 30, 1983): 7–17; "Intel's Founding," Virtual Vault (Santa Clara, CA: Intel), accessed September 30, 2021, https://www.intel.com/content/www/us/en/history/virtual-vault/articles/intels-founding.html.

3   Kim, "Interfirm Cooperation," 780.

4   William Aspray, "Interview with Tadashi Sasaki" (IEEE History Center, May 25, 1994), sec. Microprocessors and Busicom, http://ethw.org/Oral-History:Tadashi_Sasaki; Bo Lojek, *History of Semiconductor Engineering* (Berlin, Heidelberg: Springer, 2007), 359–360.

5   Leslie Berlin, *The Man Behind the Microchip: Robert Noyce and the Invention of Silicon Valley* (New York: Oxford University Press, 2005), 184.

6   Berlin, *Man Behind the Microchip*, 103–104.

7   Aspray, "Interview with Tadashi Sasaki," sec. Microprocessors and Busicom.

8   Yukio Shimura, " 'Rocket Sasaki' and His Calculator Strategy," Mr. Shimura's Essays with Historic Photos (Tokyo: Society of Semiconductor Industry Specialists, September 2011), https://www.shmj.or.jp/shimura/shimura_E/ssis_shimura2_10E.html.

9   Taro Tanimitsu, *The Origin of Semiconductor Industry in Japan : Fundamental Reasons Why Only Japan, except USA, Could Establish Strong Semiconductor Industry?* (Yamaguchi City, Japan: Economic Society of Yamaguchi University, 1996), 39; William C. Hittinger, "Metal-Oxide-Semiconductor Technology," *Scientific American* 229, no. 2 (September 30, 1973): 48–59.

10   Tout, "Sharp QT-8D."

11   Takashi Sugimoto and Tomomi Kikuchi, "Obituary: 'Rocket Sasaki,' Executive at Sharp and Mentor to Masayoshi Son," Nikkei Asia (Nikkei, February 15, 2018), https://asia.nikkei.com/Business/Obituary-Rocket-Sasaki-executive-at-Sharp-and-mentor-to-Masayoshi-Son.

12   Lojek, *History of Semiconductor Engineering*, 359–360.

13   Aspray, "Interview with Tadashi Sasaki," sec. Microprocessors and Busicom.

14   Kadokura, "Chronology of Japanese Calculating Machines"; "Operator's Manual for Busicom HL-21 Hand-Operated Calculator" (Tokyo: Busicom Computer Corporation, n.d.), 1.

15   "Manual for Busicom HL-21."

16   Serge Devidts, "Busicom 161 (Version-1)," Calcuseum, June 19, 2020, http://www.calcuseum.com/SCRAPBOOK/BONUS/44984/1.htm;   Rick Bensene, "Nippon Computing Machine Co. (NCM) Busicom 202 Desktop Calculator," The Old Calculator Web Museum, accessed October 1, 2021, https://www.oldcalculatormuseum.com/w-bus202.html; James N. Carr and Stephen T. McClellan, "The Business Machines Industry in Japan 1962-1967," *Overseas Business Reports* (Washington: U.S. Department of Commerce, 1968), 5.

17   "ビジコン デスクトップ電卓 [Busicom Desktop Calculator]," Dentaku Museum, accessed September 24, 2021, http://www.dentaku-museum.com/calc/calc/10-busicom/busicomd/busicomd.html.

18   Nigel Tout, "Busicom Junior," Vintage Calculators Web Museum, accessed October 1, 2021, http://www.vintagecalculators.com/html/ncr_18-16_-_busicom_junior.html; Serge Devidts, "National Cash Register (NCR) 18-15," Calcuseum, June 2017, http://www.calcuseum.com/SCRAPBOOK/BONUS/15153/1.htm.

19   J. F. Nolan, "Survey of Electronic Displays," *SAE Transactions* 84 (October 1, 1975): 936–949.

20   Nigel Tout, "Calculator Displays," Vintage Calculators Web Museum, accessed November 1, 2021, http://www.vintagecalculators.com/html/calculator_displays.html; "Electronics Abroad," *Electronics* (New York: McGraw Hill, May 1967).

21   William Aspray, "The Intel 4004 Microprocessor: What Constituted Invention?," *IEEE Annals of the History of Computing* 19, no. 3 (1997): 7–8, https://doi.org/10.1109/85.601727.

22   Aspray, "Intel 4004, " 7–8.

23   "Single Chip Calculator Hits the Finish Line," *Electronics* (New York: McGraw Hill, February 1971).

24   Rick Bensene, "NCR 18-15 Desktop Calculator," Old Calculator Museum, accessed October 1, 2021, https://www.oldcalculatormuseum.com/w-ncr18-15.html; Tout, "Busicom Junior."

25   "ビジコン ポケット電卓 [Busicom Pocket Calculator]," Dentaku

Museum, accessed October 3, 2021, http://www.dentaku-museum.com/calc/calc/10-busicom/busicom/busicom.html.

26 Nigel Tout, "Busicom LE-120A and LE-120S," Vintage Calculators Web Museum, accessed October 3, 2021, http://www.vintagecalculators.com/html/busicom_le-120a_-_le-120s.html; "Busicom Pocket Calculator."

27 "Busicom Pocket Calculator"; "Historical Rates - JPY to USD, January 1, 1970, to February 1, 1970," fxtop.com, accessed October 3, 2021.

28 Tout, "Busicom LE-120A and LE-120S"; "Busicom Pocket Calculator."

29 "Busicom Pocket Calculator."

30 Federico Faggin et al., "Oral History Panel on the Development and Promotion of the Intel 4004 Microprocessor," 2007, 4, 6, https://www.computerhistory.org/collections/catalog/102658187; Aspray, "Intel 4004," 7–8.

31 Faggin et al., "Oral History Panel on the Intel 4004," 4.

32 Faggin et al., "Oral History Panel, " 7.

33 Aspray, "Interview with Tadashi Sasaki," sec. Microprocessors and Busicom.

34 Aspray, "Interview with Tadashi Sasaki," sec. Microprocessors and Busicom; Aspray, "Intel 4004," 4.

35 Aspray, "Interview with Tadashi Sasaki," sec. Microprocessors and Busicom.

36 Ken Shirriff, "The Surprising Story of the First Microprocessors," *IEEE Spectrum* 53, no. 9 (2016): 48–54, https://doi.org/10.1109/MSPEC.2016.7551353.

37 Aspray, "Intel 4004," 8–9.

38 Berlin, *Man Behind the Microchip*, 185–186; Lojek, *History of Semiconductor Engineering*, 360.

39 Masatoshi Shima, "The 4004 CPU of My Youth," *IEEE Solid-State Circuits Magazine* 1, no. 1 (2009): 41, https://doi.org/10.1109/MSSC.2008.930946.

40 Aspray, "Intel 4004," 9.

41 Berlin, *Man Behind the Microchip*, 185–186.

42 Berlin, *Man Behind the Microchip*, 187–188; Faggin et al., "Oral History Panel on the Intel 4004," 6–10.

43 Berlin, *Man Behind the Microchip*, 187–188; Faggin et al., "Oral History Panel," 10.

44   Federico Faggin, "The Making of the First Microprocessor," *IEEE Solid-State Circuits Magazine* 1, no. 1 (2009): 10–12, https://doi.org/10.1109/MSSC.2008.930938; Faggin et al., "Oral History Panel," 10.

45   Faggin, "Making of the First Microprocessor," 12.

46   Faggin, "Making of the First Microprocessor," 13, 17.

47   Faggin, "Making of the First Microprocessor," 14–15.

48   Faggin, "Making of the First Microprocessor," 16; "MCS-4 Micro Computer Set Users Manual" (Santa Clara, CA: Intel, March 1973), 4.

49   "The Story of the Intel® 4004" (Santa Clara, CA: Intel), accessed October 8, 2021, https://www.intel.co.uk/content/www/uk/en/history/museum-story-of-intel-4004.html.

50   Faggin et al., "Oral History Panel," 18–19.

51   "Busicom 141-PF," IPSJ Computer Museum (Information Processing Society of Japan), accessed October 8, 2021, https://museum.ipsj.or.jp/en/heritage/Busicom_141-PF.html.

52   Andy Kessler, "The Chip That Changed the World," *Wall Street Journal*, 2021, https://www.wsj.com/articles/the-chip-that-changed-the-world-microprocessor-computing-transistor-breakthrough-intel-11636903999; "Busicom Catalogue," n.d., 11.

53   Fields et al., "U.S. Firms Gird for Calculator Battle," 83.

54   Nicholas Valéry, "Shopping Around for a Calculator," *New Scientist*, May 31, 1973, 549.

55   Gerald M. Walker and Charles Cohen, "Japanese Electronics Industry Rebounds after Two Years of Adversity," *Electronics* (New York: McGraw Hill, November 20, 1972), 92.

56   Fields et al., "U.S. Firms Gird for Calculator Battle," 84.

57   Marilyn Offenheiser, "U.S. Homes in on Calculators," *Electronics* 45, no. 20 (September 1972): 69–72.

58   Joerg Woerner, "Texas Instruments TI-2500 / Datamath Version 1," Datamath Calculator Museum, December 5, 2001, http://www.datamath.org/BASIC/DATAMATH/ti-2500-1.htm.

59   Nicholas Valéry, "The Electronic Slide-Rule Comes of Age," *New Scientist*, February 27, 1975.

60   Berlin, *Man Behind the Microchip*, 196.

61   "Announcing a New Era of Integrated Electronics: The Intel 4004," Virtual Vault (Santa Clara, CA: Intel), accessed October 12, 2021, https://www

.intel.com/content/www/us/en/history/virtual-vault/articles/the-intel-4004
.html; Stephen Cass, "Chip Hall of Fame: Intel 4004 Microprocessor," IEEE
Spectrum, July 2, 2018.

62   Nigel Tout, "Busicom / Nippon Calculating Machine Corp," Vintage Calculators Web Museum, accessed October 12, 2021, http://www
.vintagecalculators.com/html/busicom_-_ncm.html.

12. THE HEWLETT-PACKARD HP-35

1   Steve Leibson, "The HP 9100 Project: An Exothermic Reaction,"
HP9825.com, June 5, 2006, sec. Cast aluminum beauty, http://www.hp9825
.com/html/the_9100_part_2.html; David S. Cochran, "A Quarter Century at
HP," HP Memory Project, 2011, sec. Scientific Calculators, circa 1966, https://
www.hpmemoryproject.org/timeline/dave_cochran/a_quarter_century_at_
hp_00.htm.

2   "Hewlett-Packard History," Silicon Valley Historical Association, 2008,
https://www.siliconvalleyhistorical.org/hewlett-packard-history; "Hewlett-
Packard Company Annual Report" (Palo Alto, CA: Hewlett-Packard, 1968),
1, 5.

3   Leibson, "The HP 9100 Project," sec. Cast aluminum beauty.

4   David G. Hicks, "HP 9100A/B," The Museum of HP Calculators, sec.
The Case: rebuilding the benchmark, accessed October 16, 2021, https://www
.hpmuseum.org/hp9100.htm.

5   Hicks, "HP 9100A/B."

6   Sheldon Edelman, "Made in USA . . . Finally!," *The Electronic Engineer*,
March 1972, http://www.vcalc.net/hp-35.htm.

7   Chuck House, "Hewlett-Packard and Personal Computing Systems," in
A History of Personal Workstations (New York, NY, USA: Association for Computing Machinery, 1988), 414, https://doi.org/10.1145/61975.66923.

8   Tout, "Sharp QT-8D"; "International Newsletter," *Electronics* (New
York: McGraw Hill, April 14, 1969), 219.

9   David G. Hicks, "HP-35," The Museum of HP Calculators, accessed
October 21, 2021, https://www.hpmuseum.org/hp35.htm.

10   Edelman, "Made in USA."

11   Gordon E. Moore, "Cramming More Components onto Integrated
Circuits, Reprinted from *Electronics*, Volume 38, Number 8, April 19, 1965,

Pp.114 Ff.," *IEEE Solid-State Circuits Society Newsletter* 11, no. 3 (2006): 33–35, https://doi.org/10.1109/N-SSC.2006.4785860.

12   Kathryn Lorimer Koken, "Packard, David (1912–1996), Industrialist and Philanthropist" *American National Biography* (Oxford University Press, 2001), https://doi.org/10.1093/anb/9780198606697.article.1302632.

13   Edward T. Liljenwall, "Packaging the Pocket Calculator," *Hewlett-Packard Journal* 23, no. 10 (1972): 12, https://www.hpl.hp.com/hpjournal/72jun/jun72.htm.

14   Cochran, "A Quarter Century at HP," sec. The HP 9100A in Bill Hewlett's Pocket, circa 1969.

15   Liljenwall, "Packaging the Pocket Calculator."

16   Liljenwall, "Packaging the Pocket Calculator."

17   Edelman, "Made in USA."

18   Edelman, "Made in USA."

19   Cochran, "A Quarter Century at HP," sec. Scientific Calculators, circa 1966.

20   Leibson, "The HP 9100 Project," sec. The HP 9100A's core.

21   Cochran, "A Quarter Century at HP," sec. The HP 9100A in Bill Hewlett's Pocket, circa 1969.

22   Solomon Fabricant, "The 'Recession' of 1969–1970," in *Economic Research: Retrospect and Prospect*, vol. 1 (National Bureau of Economic Research, 1972), 89–136.

23   Cochran, "A Quarter Century at HP," sec. The HP 9100A in Bill Hewlett's Pocket, circa 1969.

24   Edelman, "Made in USA."

25   Thomas M. Whitney, France Rodé, and Chung C. Tung, "The 'Powerful Pocketful': An Electronic Calculator Challenges the Slide Rule," *Hewlett-Packard Journal* 23, no. 10 (1972): 4, https://www.hpl.hp.com/hpjournal/72jun/jun72.htm.

26   Edelman, "Made in USA."

27   Jacques Laporte, "HP 35 ROM Step by Step," accessed October 22, 2021, https://www.jacques-laporte.org/HP3520ROM.htm.

28   Cochran, "A Quarter Century at HP," sec. The HP 9100A in Bill Hewlett's Pocket, circa 1969.

29   Boyer, *History of Mathematics*, 15–16.

30   David Fowler and Eleanor Robson, "Square Root Approximations in

Old Babylonian Mathematics: YBC 7289 in Context," *Historia Mathematica* 25, no. 4 (November 1, 1998): 366, https://doi.org/10.1006/HMAT.1998 .2209; Boyer, *History of Mathematics*, 31.

31   S. L. Zabell, "Alan Turing and the Central Limit Theorem," *The American Mathematical Monthly* 102, no. 6 (October 8, 1995): 484, https://doi .org/10.2307/2974762.

32   Turing, "On Computable Numbers."

33   Andy Matuschal and Michael A. Nielsen, "Quantum Computing for the Very Curious" (San Francisco, CA, 2019), https://quantum.country/ qcvc.

34   Leibson, "The HP 9100 Project," sec. Too much static; Cochran, "A Quarter Century at HP," sec. Scientific Calculators, circa 1966.

35   David S Cochran, "Internal Programming of the 9100A Calculator," *Hewlett-Packard Journal* 20, no. 1 (1968): 14–16.

36   Jack E. Volder, "The Birth of CORDIC," *Journal of VLSI Signal Processing Systems for Signal, Image and Video Technology 2000* 25, no. 2 (June 1, 2000): 101–102, https://doi.org/10.1023/A:1008110704586; Bill Gunston, "General Dynamics B-58 Hustler," in *Bombers of the West* (London: Allan, 1973), 185–213.

37   Jack Volder, "The CORDIC Computing Technique," in *Papers Presented at the March 3–5, 1959, Western Joint Computer Conference*, IRE-AIEE-ACM '59 (Western) (New York, NY, USA: Association for Computing Machinery, 1959), 257, https://doi.org/10.1145/1457838.1457886.

38   Volder, "Birth of CORDIC," 102.

39   Volder, "CORDIC Computing Technique," 257.

40   Leibson, "The HP 9100 Project," sec. CORDIC and the Hustler.

41   Dave Cochran, "First-Hand: Origins of Hewlett Packard 35 (HP-35)" (IEEE History Center), accessed October 28, 2021, https://ethw.org/First -Hand:Origins_of_Hewlett_Packard_35_(HP-35).

42   Leibson, "The HP 9100 Project," sec. Cast aluminum beauty.

43   "HP-35 Operating Manual" (Cupertino, CA: Hewlett-Packard, n.d.), i.

44   David S Cochran, "Algorithms and Accuracy in the HP-35," *Hewlett-Packard Journal* 23, no. 10 (1972): 10, https://www.hpl.hp.com/ hpjournal/72jun/jun72.htm.

45   Hicks, "HP-35," sec. The Manual.

46   Hicks, "HP-35," sec. Introduction.

47   "CPI Inflation Calculator."

48   Jim Hughes, "The HP-35," Codex99, May 2, 2018, sec. Market It And They Will Come, Or, Maybe Not, http://codex99.com/design/the-hp35 .html.

49   John Minck, "Inside HP: A Narrative History Of Hewlett-Packard From 1939–1990," November 2011, 82, https://www.hpmemoryproject.org/ timeline/john_minck/HPNarrative190505.pdf.

50   Hughes, "The HP-35," sec. Market It And They Will Come, Or, Maybe Not.

51   Minck, "Inside HP," 82; Dejan Ristanović and Jelica Protić, "Once upon a Pocket: Programmable Calculators from the Late 1970s and Early 1980s and the Social Networks around Them," *IEEE Annals of the History of Computing* 34, no. 3 (2012): 57, https://doi.org/10.1109/MAHC .2011.63.

52   Michael S. Malone, *Bill & Dave: How Hewlett and Packard Built the World's Greatest Company* (New York: Portfolio, 2008), 272.

53   "In 1972 . . . the First Scientific Hand-Held Calculator (HP-35) Is Introduced for $395" (Hewlett-Packard), accessed October 29, 2021, www.hp .com/calculators; Bill Doolittle, "Mission to the Middle Kingdom," *Measure* (Hewlett-Packard, December 1972), 3; "Calculator, Pocket, Electronic, HP-35," National Air and Space Museum, accessed October 29, 2021, https:// airandspace.si.edu/collection-objects/calculator-pocket-electronic-hp-35/ nasm_A19850582000; Malone, *Bill & Dave*, 271.

54   Malone, 273; Minck, "Inside HP," 53.

55   James, "The Deaths of the Slide Rule," 6–9; Lee Dembart, "Slide Rule Going the Way of Abacus as Pocket Calculator Moves In," *New York Times*, October 10, 1977, https://www.nytimes.com/1977/10/10/archives/slide-rule -going-the-way-of-abacus-as-pocket-calculator-moves-in.html.

56   Minck, "Inside HP," 83; Malone, *Bill & Dave*, 271–273.

13. THE PULSAR TIME COMPUTER CALCULATOR

1   *The Tonight Show Starring Johnny Carson* (NBC, May 6, 1970).

2   Joe Thompson, "The Lost Chapter: A Concise History of the LED Watch," Hodinkee, February 26, 2018, https://www.hodinkee.com/articles/ four-revolutions-led-watches.

3   Eric Bruton, *The History of Clocks and Watches* (London: Little, Brown, 2000), 32–33.

4   Bruton, *History of Clocks and Watches*, 48, 68–69.

5   Bruton, *History of Clocks and Watches*, 109–110; Carlene Stephens and Maggie Dennis, "Engineering Time: Inventing the Electronic Wristwatch," *The British Journal for the History of Science* 33, no. 4 (2000): 478.

6   Bruton, *History of Clocks and Watches*, 183–198.

7   "The Watch: Its Invention and History," *Scientific American* 37, no. 6 (November 9, 1877): 89.

8   Bruton, *History of Clocks and Watches*, 195–198; Stephens and Dennis, "Engineering Time," 480–482.

9   Stephens and Dennis, 481–482; Warren A. Marrison, "The Evolution of the Quartz Crystal Clock," *Bell System Technical Journal* 27, no. 3 (1948): 523–531, https://doi.org/10.1002/J.1538-7305.1948.TB01343.X.

10   Lance Day and Ian McNeil, "Hetzel, Max, " *Biographical Dictionary of the History of Technology* (New York: Routledge, 2002).

11   Stephens and Dennis, "Engineering Time," 480; Jarett Harkness, "Hamilton Electric: The Race to Create the World's First Battery-Powered Watch," Worn & Wound, May 31, 2018, https://wornandwound.com/hamilton -electric-the-race-to-create-the-worlds-first-battery-powered-watch/.

12   Harkness, "Hamilton Electric."

13   Stephens and Dennis, "Engineering Time," 492.

14   David Voss et al., "March 1880: The Curie Brothers Discover Piezoelectricity," American Physical Society, 2014.

15   Marrison, "Evolution of the Quartz Crystal Clock," 534–536.

16   Stephens and Dennis, "Engineering Time," 492.

17   "Seiko Quartz Astron 35SQ" (Nagano: Epson), accessed November 11, 2021, https://global.epson.com/company/corporate_history/milestone_ products/pdf/05_35sq.pdf.

18   Alex Newson, *Fifty Watches That Changed the World*, Design Museum Fifty (Octopus, 2015), 40; Thompson, "History of the LED Watch."

19   Thompson, "History of the LED Watch."

20   *Pulsar Product Description* (Hamilton Watch Company, 1972), https:// www.youtube.com/watch?v=a5szJYA_z44.

21   Stephens and Dennis, "Engineering Time," 492.

22   Thompson, "History of the LED Watch."

23 Thompson; Stephens and Dennis, "Engineering Time," 493.

24 Thompson, "History of the LED Watch"; Charlie Burton, "James Bond's Most Groundbreaking Watch Was a Hamilton," *GQ*, March 2, 2020, https://www.gq-magazine.co.uk/watches/article/hamilton-pulsar-p2-2900-james-bond; Maxine Cheshire, "Ford: Pulsar's Better Idea?," *Washington Post*, December 1974; Alex Doak, "Hamilton's Time Computer, the Original Smartwatch, Is Back," Wired, November 1, 2020, https://www.wired.co.uk/article/hamilton-psr-limited-edition-watch.

25 "Pulsar Jeweler's Technical Manual" (Lancaster, PA: Time Computer, Inc., n.d.), 27; Doak, "Original Smartwatch"; Thompson, "History of the LED Watch."

26 "Novus National Semiconductor LED 1975," accessed November 12, 2021, http://www.crazywatches.pl/novus-national-semiconductor-led-1975; "TI's First LED Digital Watch," Texas Instruments Records (SMU Libraries Digital Collections), accessed November 12, 2021, https://digitalcollections.smu.edu/digital/collection/tir/id/205/rec/2; "Timeband Fairchild LED 1975," accessed November 12, 2021, http://www.crazywatches.pl/timeband-fairchild-led-1975.

27 Thompson, "History of the LED Watch."

28 "Omega 1600 TC1 Time Computer LED 1973," accessed November 12, 2021, http://www.crazywatches.pl/omega-1600-tc1-time-computer-led-1973.

29 "Electronics Review," *Electronics* 48, no. 16 (August 1975): 48.

30 "Pulsar Jeweler's Technical Manual," 27, 32, 34.

31 *Wish Book for the 1975 Christmas Season* (Chicago: Sears, Roebuck and Company, 1975), 187.

32 "Busicom Pocket Calculator."

33 Chung C. Tung, "The 'Personal Computer': A Fully Programmable Pocket Calculator," *Hewlett-Packard Journal* 25, no. 9 (1974): 2–7, https://www.hpl.hp.com/hpjournal/pdfs/IssuePDFs/1974-05.pdf.

34 Nigel Tout, "Casio Mini," Vintage Calculators Web Museum, accessed November 12, 2021, http://www.vintagecalculators.com/html/casio_mini.html; "$60 Calculator," *Popular Science* 201, no. 5 (November 1972): 107.

35 Nigel Tout, "Clive Sinclair and the Pocket Calculator," Vintage Calculators Web Museum, 2021, http://www.vintagecalculators.com/html/sinclair_-_the_pocket_calculator.html; "Small Pocket Calculator," *Wireless World*, August 1972.

36   "Historical Rates - GBP to USD, August 1972," fxtop.com, accessed November 12, 2021.

37   Nicola Slawson, "Paul Newman's Rolex Watch Sells for Record $17.8m," *Guardian*, October 28, 2017, https://www.theguardian.com/film/2017/oct/28/paul-newman-rolex-watch-sells-record-178m-auction; Rebecca Maksel, "The Watches That Went to the Moon," *Air & Space Magazine* (National Air and Space Museum, December 2015), https://www.airspacemag.com/space/space-timekeepers-180957295/.

38   Thompson, "History of the LED Watch"; Guy Ball, "The Pulsar Calculator Watch," ed. Nigel Tout, The International Calculator Collector, accessed November 18, 2021, http://www.vintagecalculators.com/html/pulsar_calculator_watch1.html; Judy Licht, "10 O' Clock News" (New York: WNEW-TV, December 1975), https://www.youtube.com/watch?v=a5szJYA_z44.

39   Licht, "10 O' Clock News."

40   Ball, "The Pulsar Calculator Watch."

41   "Pulsar Calculator Time Computer LED 1975," accessed November 18, 2021, http://www.crazywatches.pl/pulsar-calculator-time-computer-led-1975; V. Elaine Smay, "A Wristwatch That Sums It Up," *Popular Science* 208, no. 4 (April 1976): 54.

42   "Pulsar Jeweler's Technical Manual," 3.

43   Ball, "The Pulsar Calculator Watch."

44   Guy Ball, "The Hughes Aircraft Company LED Modules and Calculator Watch," LED Watches, 2004, https://web.archive.org/web/20150819061625/http://www.ledwatches.net/articles/Hughes Aircraft Company's Calculator Watch.htm; "Compuchron Hughes Aircraft LED 1973," accessed November 12, 2021, http://www.crazywatches.pl/compuchron-hughes-aircraft-led-1973; Adam Harras, "Hughes Aircraft (Aka Timeulator)," Digital Watch, accessed November 18, 2021, http://www.digital-watch.com/DWL/1work/hughes-calculator-watch-1977.

45   David G. Hicks, "HP-01," The Museum of HP Calculators, accessed November 18, 2021, https://www.hpmuseum.org/hp01.htm.

46   "Calculator Prices Continue to Plummet," *New Scientist*, June 6, 1974.

47   "Demand for Calculator Chips Exceeds Supply," *Popular Electronics* (New York: Ziff-Davis, September 1974); Nicholas Valéry, "Coming of Age in the Calculator Business," *New Scientist*, November 13, 1975.

48   "'Scary Boom' in US Electronics," *New Scientist*, November 1, 1973.

49   "Calculator Prices Continue to Plummet"; "Historical Rates - GBP to USD, 25 Dec 1974," fxtop.com, accessed November 19, 2021, https://fxtop .com.

50   "Arab Oil Embargo," *Encyclopaedia Britannica* (Chicago: Encyclopaedia Britannica, October 1, 2020), https://www.britannica.com/event/ Arab-oil-embargo/; Department of Economic and Social Affairs, "World Economic Survey, 1975" (New York: United Nations, 1976), sec. Instability in the first half of the 1970s; Victor Zarnowitz and Geoffrey H. Moore, "The Recession and Recovery of 1973–1976," *Explorations in Economic Research* 4, no. 4 (1977): sec. The 1973–1975 Recession.

51   "Rockwell Shuffles Microelectronics Heads," *Electronics* 48, no. 16 (August 1975): 54.

52   "People," *Electronics* (New York: McGraw Hill, August 7, 1975).

53   Nigel Tout, "Adler Lady and Sir," Vintage Calculators Web Museum, accessed November 22, 2021, http://www.vintagecalculators.com/html/ adler_lady_-_sir.html.

54   Colin Riches, "Production Lines," *Practical Wireless* (London: IPC Magazines, July 1975), 230.

55   Joerg Woerner, "Texas Instruments Spirit of '76," Datamath Calculator Museum, December 2001, http://www.datamath.org/BASIC/TI-1200/Spirit .htm.

56   "'Compact,' 'Diary,' Checkbook All Count," *Electronics* (New York: McGraw Hill, April 3, 1975).

57   "Pierre Cardin Calculator Clock Radio," *The Pocket Calculator Show*, October 13, 2002, https://www.pocketcalculatorshow.com/magicalgadget/ calculator/pierre-cardin-calculator-clock-radio/.

58   Nigel Tout, "Hanimex Calculator-Recorder," Vintage Calculators Web Museum, accessed November 23, 2021, http://www.vintagecalculators.com/ html/hanimex_calculator-recorder.html.

59   Nathan Zeldes, "A Curious Hybrid," History of Computing, 2005, https://www.nzeldes.com/HOC/HybridCalc.htm.

60   "'Compact,' 'Diary,' Checkbook All Count"; Joerg Woerner, "Texas Instruments TI-1260," Datamath Calculator Museum, December 2001, http://www.datamath.org/BASIC/TI-1200/TI-1260.htm.

61   David G. Hicks, "HP-12C," The Museum of HP Calculators, accessed November 25, 2021, https://www.hpmuseum.org/hp12c.htm.

62   Joerg Woerner, "USMC Harrier Calculator by Texas Instruments," Datamath Calculator Museum, December 5, 2001, http://www.datamath .org/Sci/WEDGE/TI-58-Harrier.htm.

63   *Navy Aviation: AV-8B Harrier Remanufacture Strategy Is Not the Most Cost-Effective Option* (Washington, DC: General Accounting Office, February 1, 1996), 2, https://apps.dtic.mil/sti/citations/ADA304930; Joerg Woerner, "Texas Instruments TI-58," Datamath Calculator Museum, December 5, 2001, http://www.datamath.org/Sci/WEDGE/TI-58.htm.

64   Nigel Tout, "Casio CQ-1 Calculator and Clock," Vintage Calculators Web Museum, accessed November 25, 2021, http://www.vintagecalculators .com/html/casio_cq-1.html.

65   "First in the World VFD Watch-Calculator - Casio CQ1-1976," Digital Watch Forum, July 4, 2017, http://www.newdwf.com/viewtopic .php?f=67&t=8562.

66   Mark Vail, *The Synthesizer* (Oxford: Oxford University Press, 2014), 277.

67   "Casio VL-Tone VL-1 Operation Manual" (CASIO, n.d.).

68   "Casio VL-Tone," Synthmuseum, accessed November 25, 2021, https:// www.synthmuseum.com/casio/casvltone01.html.

69   Colin Marshall, "When Kraftwerk Issued Their Own Pocket Calculator Synthesizer—to Play Their Song 'Pocket Calculator' (1981)," Open Culture, June 6, 2019, https://www.openculture.com/2019/06/when-kraftwerk-issued -their-own-pocket-calculator-synthesizer.html.

70   "Casio FX-190 and FX-191," Curtamania, January 25, 2017, http://www .curtamania.com/curta/database/brand/casio/Casio FX-190 and FX-191/ index.html; "Casio Fx-190 Operation Manual" (CASIO, n.d.).

71   Nigel Tout, "Casio Mini," Vintage Calculators Web Museum, http:// www.vintagecalculators.com/html/casio_mini.html

72   Serge Devidts, "Casio QD100 (Version-1) BK," Calcuseum, December 24, 2019, http://www.calcuseum.com/SCRAPBOOK/BONUS/75146/1 .htm; Serge Devidts, "Casio SG12," Calcuseum, July 15, 2021, http://www .calcuseum.com/SCRAPBOOK/BONUS/33281/1.htm.

73   Serge Devidts, "Casio QL-10," Calcuseum, February 2017, http://www .calcuseum.com/SCRAPBOOK/BONUS/08879/1.htm.

74   Serge Devidts, "Brand/Mark - Quantity of Models," Calcuseum, November 25, 2021, http://www.calcuseum.com/PDF/smd50496.pdf; Tout, "Casio Mini."

75   Gerald R. Patton, "Throw-Away Digitals You'll Want to Keep," *Popular Mechanics* (New York: Hearst, June 1977).

76   Stephens and Dennis, "Engineering Time," 495.

77   Pieter Doensen, "Dynamic Scattering LCD," Watch: History of the Modern Wrist Watch, 2007, https://doensen.home.xs4all.nl/l1.html; "Seiko Quartz LC V.F.A. 06LC" (Nagano: Epson), accessed November 26, 2021, https://global.epson.com/company/corporate_history/milestone_products/pdf/08_06lc.pdf.

78   Thompson, "A History of the LED Watch."

14. THE TEXAS INSTRUMENTS TI-81

1   "1980 U.S. Industrial Outlook" (Washington, DC: U.S. Department of Commerce, 1980), 254–255; "1981 U.S. Industrial Outlook" (Washington, DC: U.S. Department of Commerce, 1981).

2   Fred Glass, "Sign of the Times: The Computer as Character in 'Tron,' 'War Games', and 'Superman III,'" *Film Quarterly* 38, no. 2 (1984): 16–27.

3   Plato, *Phaedrus*, trans. Harold N. Fowler, Plato in Twelve Volumes (Cambridge, MA: Harvard University Press, 1925), sec. 274de, http://data.perseus.org/texts/urn:cts:greekLit:tlg0059.tlg012.perseus-eng1.

4   "The Intellectual Effects of Electricity," *The Spectator*, November 9, 1889; "The Telephone Unmasked," *New York Times*, October 13, 1877, https://www.nytimes.com/1877/10/13/archives/the-telephone-unmasked.html.

5   Needham and Wang, "Mathematics," 70.

6   Reynolds, "Algorists vs. Abacists," 220–221.

7   Johnston, "Making the Arithmometer Count," sec. Promotion and competition.

8   "Marilyn N. Suydam," National Council of Teachers of Mathematics (Reston, VA), accessed December 5, 2021, https://www.nctm.org/Grants-and-Awards/Lifetime-Achievement-Award/Marilyn-N_-Suydam/; Marilyn Suydam, "Electronic Hand Calculators: The Implications for Pre-College Education. Final Report." (Washington, DC: National Science Foundation, February 1976).

9   Suydam, "Electronic Hand Calculators," 22.

10   Max Bell, "Needed R&D on Hand-Held Calculators," *Educa-*

tional Researcher 6, no. 5 (November 30, 2016): 8, https://doi.org/10 .3102/0013189X006005007.

11   Suydam, "Electronic Hand Calculators," 20–21.

12   David Klein, "A Brief History of American K-12 Mathematics Education in the 20th Century," in *Mathematical Cognition: A Volume in Current Perspectives on Cognition, Learning, and Instruction*, 2003, sec. Historical Outline: 1920 to 1980.

13   Bell, "Needed R&D on Hand-Held Calculators," 8–9.

14   Suydam, "Electronic Hand Calculators," 16.

15   "Bert Waits Obituary," *The Columbus Dispatch*, August 2, 2014, https:// www.legacy.com/us/obituaries/dispatch/name/bert-waits-obituary?id= 21459645; "Franklin Demana," *The Columbus Dispatch*, October 1, 2021, https://www.dispatch.com/obituaries/b0055559.

16   Joan Leitzel and Bert Waits, "Hand-Held Calculators in the Freshman Mathematics Classroom," *The American Mathematical Monthly* 83, no. 9 (1976): 731–732.

17   Leitzel and Waits, "Hand-Held Calculators, " 731–732.

18   "Texas Instruments SR-50 Scientific Calculator," *Electronics* (New York: McGraw Hill, October 3, 1974).

19   "MyCalcDB," accessed December 9, 2021, http://mycalcdb.free.fr/main .php?l=0&p=1.

20   Joerg Woerner, "Texas Instruments Little Professor (1976)," Datamath Calculator Museum, December 5, 2001, http://www.datamath.org/Edu/ Professor-76.htm; Joerg Woerner, "Texas Instruments MATH MAGIC," Datamath Calculator Museum, December 5, 2001, http://www.datamath .org/Edu/MathMagic.htm; Joerg Woerner, "Texas Instruments WIZ-A-TRON," Datamath Calculator Museum, December 5, 2001, http://www .datamath.org/Edu/WizATron.htm.

21   Gerald Luecke, Calculator with interchangeable keyset (USA, issued May 1978), http://www.google.com/patents/US4092527A.

22   Joerg Woerner, "Texas Instruments TI-1766 (1st Design)," Datamath Calculator Museum, December 2002, http://www.datamath.org/BASIC/ LCD_Classic/TI-1766_1.htm.

23   *Texas Instruments Calculators for Students (CB-272)* (Dallas, TX: Texas Instruments, 1977), http://www.datamath.net/Leaflets/CB-272_US.pdf.

24   Samuel L. Greitzer, "The Second U.S.A. Mathematical Olympiad,"

The Mathematics Teacher 67, no. 2 (1974): 115–119; Samuel L. Greitzer, "The Third U.S.A. Mathematical Olympiad," *The Mathematics Teacher* 68, no. 1 (1975): 4–9.

25 Suydam, "Electronic Hand Calculators," 78–87.

26 Sarah Banks, "A Historical Analysis of Attitudes Toward the Use of Calculators in Junior High and High School Math Classrooms in the United States Since 1975," *Master of Education Research Theses* (Cedarville University, 2011), 59, https://doi.org/10.15385/tmed.2011.1.

27 *An Agenda for Action: Recommendations for School Mathematics of the 1980s* (Reston, VA: National Council of Teachers of Mathematics, 1980), 1.

28 Charlotte Libov, "State Adding a Tool for 8th Grade Math," *New York Times*, August 31, 1986, https://www.nytimes.com/1986/08/31/nyregion/state-adding-a-tool-for-8th-grade-math.html.

29 "Calculators Allowed for Math Regents," *New York Times*, August 8, 1992, https://www.nytimes.com/1992/08/08/nyregion/calculators-allowed-for-math-regents.html; "Chicago Provides Free Calculators to Students," *New York Times*, January 5, 1988, https://www.nytimes.com/1988/01/05/science/chicago-provides-free-calculators-to-students.html.

30 Banks, "Attitudes Toward the Use of Calculators," 23, 66.

31 Bert Waits and Franklin Demana, "The Calculator and Computer Precalculus Project (C2PC): What Have We Learned in Ten Years?" *Hand-Held Technology in Mathematics and Science Education: A Collection of Papers*, 1994, 2.

32 Eric W. Weisstein, "Calculus," *MathWorld* (Wolfram Research, Inc.), accessed December 12, 2021, https://mathworld.wolfram.com/Calculus.html.

33 C. A. Browning, "The Computer/Calculator Precalculus (C2PC) Project and Levels of Graphical Understanding," in *Proceedings of the Conference on Technology in Collegiate Mathematics*, 1989, 114.

34 Joerg Woerner, "Casio Fx-7000G," Datamath Calculator Museum, 2004, http://www.datamath.org/Related/Casio/fx-7000G.htm.

35 "Pocket Calculator," *Radio-Electronics* (Farmingdale, NY: Gernsback, August 1986); "The World's First Graphic Display Programmable Scientific Calculator," *New Scientist*, October 24, 1985; "The First Interactive Graphic Calculator," *Electronics* (New York: McGraw Hill, March 20, 1975).

36 Waits and Demana, "(C2PC): What Have We Learned in Ten Years?" 2.

37 Browning, "(C2PC) Project," 117.

38   Commission on Standards for School Mathematics, *Curriculum and Evaluation Standards for School Mathematics* (Reston, VA: National Council of Teachers of Mathematics, 1989), 123.

39   David G. Hicks, "HP-28C/S," The Museum of HP Calculators, accessed December 16, 2021, https://www.hpmuseum.org/hp28c.htm.

40   Yves Nievergelt, "The Chip with the College Education: The HP-28C," *The American Mathematical Monthly* 94, no. 9 (December 16, 1987): 895–902.

41   Woerner, "Casio Fx-7000G"; Joerg Woerner, "Texas Instruments TI-81," Datamath Calculator Museum, January 19, 2003, http://www.datamath.org/Graphing/TI-81_I0990.htm; Calculator Culture, *TI-81 - Texas Instruments' First Graphing Calculator* (Australia, 2021), https://www.youtube.com/watch?v=vqEmG9iaAbw.

42   Woerner, "Casio Fx-7000G."

43   "Chip Hall of Fame: Zilog Z80 Microprocessor," *IEEE Spectrum*, June 2017.

44   Denis Brusseaux, Mehdi El Kanafi, and Nicolas Courcier, *Metal Gear Solid: Hideo Kojima's Magnum Opus* (Toulouse: Third Editions, 2019), chap. Hideo Kojima; Mark J. P. Wolf and Bernard Perron, *The Video Game Theory Reader 2* (London: Routledge, 2009), 172; Nathan Altice, *I Am Error: The Nintendo Family Computer/Entertainment System Platform* (Cambridge, MA: MIT Press, 2017), 13–14, 301.

45   Geoffrey Akst, "Graphing Calculators: A Conversation with Bert Waits," *Journal of Developmental Education* 18, no. 3 (1995): 18; *TI-81 Guidebook* (Lubbock, TX: Texas Instruments, 1990), i.

46   "Back Matter," *The College Mathematics Journal* 23, no. 4 (December 17, 1992); Waits and Demana, "(C2PC): What Have We Learned in Ten Years?" 2.

47   G. D. Foley, "The Texas Instruments TI-92 as a Vehicle for the Teaching and Learning of Function, Graphs, and Analytical Geometry," *International Journal of Computer Algebra in Mathematics Education* 4 (1997): 221.

48   Edward Laughbaum, "Working Group 3: Continued Professional Development," in *Electronic Proceedings of the Fifth International Conference on Technology in Mathematics Teaching*, ed. Manfred Borovcnik and Hermann Kautschitsch (Klagenfurt, 2001), http://wwwg.uni-klu.ac.at/stochastik.schule/ICTMT_5/ICTMT_5_CD/Working groups/Working Group 3.htm; Dale Philbrick, personal correspondence with the author, February 17, 2022.

49   Gail Burrill et al., "Handheld Graphing Technology in Secondary Mathematics," *Texas Instruments*, 2002; Paul Laumakis and Marlena Herman, "The Effect of a Calculator Training Workshop for High School Teachers on Their Students' Performance on Florida State-Wide Assessments," *International Journal for Technology in Mathematics Education* 15, no. 3 (2008); Kristina K. Hill, Ali Bicer, and Robert M. Capraro, "Effect of Teachers' Professional Development from Mathforward™ on Students' Math Achievement," *International Journal of Research in Education and Science* 3, no. 1 (2017): 67–74.

50   "Research Library" (Texas Instruments), accessed December 18, 2021, http://ti-researchlibrary.com/default.aspx.

51   Andrew Trotter, "After Four Decades, Pioneer of Calculator Still Leads K–12 Field," EducationWeek, October 16, 2007, https://www.edweek.org/policy-politics/after-four-decades-pioneer-of-calculator-still-leads-k-12-field/2007/10.

52   "Client Profile: Texas Instruments (2007)," OpenSecrets, accessed December 22, 2021, https://www.opensecrets.org/federal-lobbying/clients/lobbyists?cycle=2007&id=D000000722.

53   Zachary Crockett, "Is the Era of the $100+ Graphing Calculator Coming to an End?" The Hustle, September 22, 2019, https://thehustle.co/graphing-calculators-expensive/; Jason Stanford, "Why Is Texas Instruments Lobbying Schools to Make Algebra II Mandatory?" Alternet, April 24, 2013, https://www.alternet.org/2013/04/why-texas-instruments-lobbying-schools-make-algebra-ii-mandatory/; Morgan Smith, "In Texas, Nixing Algebra II Not Out of the Equation," *Texas Tribune*, April 12, 2013, https://www.texastribune.org/2013/04/12/texas-nixing-algebra-ii-not-out-equation/.

54   "T³: Our Mission," Education Technology: Professional Development (Texas Instruments), accessed December 20, 2021, https://education.ti.com/en/professional-development/t3-our-mission; "About T³," T³ Learns, accessed December 20, 2021, http://www.t3learns.org/about-t3/.

55   "SAT Calculator Policy," SAT Suite of Assessments (College Board and National Merit Scholarship Corporation), accessed December 20, 2021, https://collegereadiness.collegeboard.org/sat/taking-the-test/calculator-policy.

56   "Texas Instruments Celebrates Shipping More than 20 Million Graphing Calculators," (press release) Texas Instruments, September 6, 2000; "TI

Has Screenagers Screaming for More With the Sale of Its 25 Millionth Graphing Handheld" (press release), Texas Instruments, July 22, 2003).

57   Matt McFarland, "The Unstoppable TI-84 Plus: How an Outdated Calculator Still Holds a Monopoly on Classrooms," *Washington Post*, September 2, 2014,   https://www.washingtonpost.com/news/innovations/wp/2014/09/02/ the-unstoppable-ti-84-plus-how-an-outdated-calculator-still-holds-a -monopoly-on-classrooms/.

58   Crockett, "Is the Era of the $100+ Graphing Calculator Coming to an End?"

59   Commission on Standards for School Mathematics, *Curriculum and Evaluation Standards for School Mathematics*, 8; Mark Clayton, "Calculators in Class: Freedom from Scratch Paper or 'Crutch'?" *Christian Science Monitor*, May 23, 2000, https://www.csmonitor.com/2000/0523/p20s1.html.

60   "Calculators in Schools: Pupils Face Limits," BBC News (BBC, 2011), https://www.bbc.co.uk/news/education-15984003.

61   Stephanie Simon, "The School Standards Rebellion," *Politico*, February 14, 2014, https://www.politico.com/story/2014/02/education-standards -reform-high-school-college-103510.

62   "SAT Calculator Policy."

63   Lilian Smith, "Opinion: Ban the Use of High-End Calculators During the SAT," *Los Angeles Times*, January 27, 2020, https://www.latimes.com/ opinion/story/2020-01-27/calculator-sat-testing-math.

64   Crockett, "Is the Era of the $100+ Graphing Calculator Coming to an End?"

65   Texas Instruments, *2011 Annual Report* (Dallas, TX: Texas Instruments, 2011), 47, https://investor.ti.com/static-files/2b690577-f4c8-4e19-80c6-5d8cba 365c80; Texas Instruments, *2020 Annual Report* (Dallas, TX: Texas Instruments, 2020), 18, https://investor.ti.com/static-files/05b7598d-4a01-4f45-a63d -058c69a165ad.

66   Aimee J. Ellington, "A Meta-Analysis of the Effects of Calculators on Students' Achievement and Attitude Levels in Precollege Mathematics Classes," *Journal for Research in Mathematics Education* 34, no. 5 (2003): 456, https://doi.org/10.2307/30034795.

67   Jeremy Hodgen et al., "Improving Mathematics in Key Stages Two and Three: Evidence Review," *Education Endowment Foundation*, 2018.

15. SOFTWARE ARTS VISICALC

1   William J. Hawkins, "Solar Calc," *Popular Science*, April 1978; Nigel Tout, "Teal Photon," Vintage Calculators Web Museum, accessed January 7, 2022, http://www.vintagecalculators.com/html/teal_photon.html; "ティール (東京電子応用研究所) [TEAL: Tokyo Electronic Application Laboratory]," Dentaku Museum, accessed January 7, 2022, http://www.dentaku-museum .com/calc/calc/17-teal/teal/teal.html.

2   Nigel Tout, "Casio Mini Card LC-78," Vintage Calculators Web Museum, accessed January 7, 2022, http://www.vintagecalculators.com/html/ casio_mini_card_lc-78.html; John Wolff, "The Casio Computer Company: Early Electronic Calculators," John Wolff's Web Museum, September 10, 2011, http://www.johnwolff.id.au/calculators/Casio/Casio.htm.

3   David G. Hicks, "HP-41C/CV/CX," The Museum of HP Calculators, accessed January 7, 2022, https://www.hpmuseum.org/hp41.htm.

4   Tracy Robnett Licklider, "Ten Years of Rows and Columns," *Byte* 14, no. 13 (December 1989): 326; Peter Reuell, "A Vision of Computing's Future," *Harvard Gazette*, March 22, 2012, https://news.harvard.edu/gazette/ story/2012/03/a-vision-of-the-computing-future/.

5   Dan Bricklin, "Tool Maker," *Computerworld* 26, no. 25 (June 1992): 24–25; Martin Campbell-Kelly and Paul Ceruzzi, "An Interview with Dan Bricklin and Bob Frankston (OH 402), Charles Babbage Institute, 2004, 11–12, https://conservancy.umn.edu/handle/11299/113026; Walter Isaacson, "Dawn of a Revolution," *Harvard Gazette*, September 20, 2013, https://news .harvard.edu/gazette/story/2013/09/dawn-of-a-revolution/.

6   E. W. Paxson, *Hand Calculator Programs for Staff Officers* (Fort Belvoir, VA: Defense Technical Information Center, 1978); Woerner, "USMC Harrier Calculator"; S. Jagannath, *Calculator Programs for the Hydrocarbon Processing Industries* (Houston: Gulf, 1980); Kenneth J. Rothman and John D. Boice, *Epidemiologic Analysis with a Programmable Calculator* (Bethesda, MD: U.S. Dept. of Health, Education, and Welfare, 1979); John B. Wright, *Computer Programs for Three-Dimensional Cable Problems in Tethered-Balloon Applications* (Hanscom AFB, MA: Air Force Geophysics Laboratories, 1977); Wayne D. Shepperd, *Hand-Held-Calculator Programs for the Field Forester* (Fort Collins, CO: Rocky Mountain Forest and Range Experiment Station, Forest Service, U.S. Dept. of Agriculture, 1980).

7   Dan Bricklin, "There Is Always New Technology," Business of Software 2010, August 22, 2011, https://businessofsoftware.org/2011/08/dan-bricklin -at-business-of-software-2010-there-is-always-new-technology-when-should -we-care-how-do-we-take-advantage-of-it-video-transcript/.

8   Reuell, "A Vision of Computing's Future."

9   "Linolex Legal Throughput Lets Lawyers Practice Law," *ABA Journal* 60 (June 1974): 693; "The Typing Crisis in Law," *ABA Journal* 59 (August 1973): 821.

10   Licklider, "Ten Years of Rows and Columns," 326; Dan Bricklin, "The Idea," Software Arts and VisiCalc, accessed January 22, 2022, http://www .bricklin.com/history/saiidea.htm.

11   Campbell-Kelly and Ceruzzi, "Interview with Dan Bricklin and Bob Frankston," 11–13.

12   Licklider, "Ten Years of Rows and Columns," 326; Dan Bricklin, "Special Short Paper for the HBS Advertising Course," December 1, 1978, 4, http://www.bricklin.com/anonymous/bricklin-1978-visicalc-paper.pdf.

13   Martin Campbell-Kelly, "The Rise and Rise of the Spreadsheet," in *The History of Mathematical Tables* (Oxford: Oxford University Press, 2003), 324, https://doi.org/10.1093/acprof:oso/9780198508410.003.0013.

14   Campbell-Kelly, "Rise and Rise, " 324–325.

15   Licklider, "Ten Years of Rows and Columns," 326.

16   Ken R. Adcock, "Comparison of Three Major Time-Sharing Financial Modeling Languages," *SIGSIM Simulation Digest* 7, no. 3 (April 1976): 18–19, https://doi.org/10.1145/1102746.1102748.

17   Campbell-Kelly, "Rise and Rise," 326–327; Yuji Ijiri, "Review of 'Simulation of the Firm Through a Budget Computer Program' by Richard Mattessich," *The Journal of Business* 38, no. 4 (1965): 430.

18   Campbell-Kelly and Ceruzzi, "Interview with Dan Bricklin and Bob Frankston," 14.

19   Bricklin, "The Idea."

20   Licklider, "Ten Years of Rows and Columns," 326; Bricklin, "The Idea."

21   Dan Fylstra, "Personal Account: The Creation and Destruction of VisiCalc," May 2004, 4; Campbell-Kelly and Ceruzzi, "Interview with Dan Bricklin and Bob Frankston," 10, 13.

22   Fylstra, "The Creation and Destruction of VisiCalc," 2, 5.

23   Campbell-Kelly and Ceruzzi, "Interview with Dan Bricklin and Bob Frankston," 15.

24   Licklider, "Ten Years of Rows and Columns," 327.

25   Licklider, "Ten Years of Rows and Columns," 326; Campbell-Kelly and Ceruzzi, "Interview with Dan Bricklin and Bob Frankston," 19–20.

26   Andrew Pollack, "How a Software Winner Went Sour," *New York Times*, February 1984, https://www.nytimes.com/1984/02/26/business/how-a-software-winner-went-sour.html.

27   Bob Frankston, "Email Message," April 15, 1999, https://dssresources.com/history/frankston4151999b.html; Fylstra, "The Creation and Destruction of VisiCalc," 8.

28   "VisiCalc (Advertisement)," *BYTE* (May 1979).

29   Fylstra, "Creation and Destruction of VisiCalc," 9–10.

30   Peter Jennings, "VisiCalc 1979 (Part 3)," Benlo Park, accessed January 20, 2022, http://www.benlo.com/visicalc/visicalc3.html; Campbell-Kelly and Ceruzzi, "Interview with Dan Bricklin and Bob Frankston," 23; Walter Isaacson, *Steve Jobs* (New York: Simon & Schuster, 2013), 80–81.

31   Leslie Haddon, "The Home Computer: The Making of a Consumer Electronic," *Science as Culture* 1, no. 2 (1988): sec. The Home Computer, https://doi.org/10.1080/09505438809526198.

32   Dan Bricklin, "Ben Rosen's Reaction," Software Arts and VisiCalc, accessed January 20, 2022, http://www.bricklin.com/history/rosenletter.htm; James W. Cortada, "How the IBM PC Won, Then Lost, the Personal Computer Market," *IEEE Spectrum*, July 21, 2021, https://spectrum.ieee.org/how-the-ibm-pc-won-then-lost-the-personal-computer-market.

33   Dan Bricklin, "National Computer Conference," Software Arts and VisiCalc, accessed January 20, 2022, http://www.bricklin.com/history/saincc.htm.

34   Fylstra, "Creation and Destruction of VisiCalc," 8–9; Jennings, "VisiCalc 1979 (Part 3)."

35   Campbell-Kelly and Ceruzzi, "Interview with Dan Bricklin and Bob Frankston," 28.

36   Campbell-Kelly and Ceruzzi, "Interview with Dan Bricklin and Bob Frankston," 21–22; Jennings, "VisiCalc 1979 (Part 3)."

37   Campbell-Kelly and Ceruzzi, "Interview with Dan Bricklin and Bob Frankston," 28.

38   Jennings, "VisiCalc 1979 (Part 3)"; Dan Bricklin, "The First Product,"

Software Arts and VisiCalc, accessed January 21, 2022, http://www.bricklin.com/history/saiproduct1.htm.

39   Licklider, "Ten Years of Rows and Columns," 327.

40   Bricklin, "Ben Rosen's Reaction."

41   Robert E. Ramsdell, "The Power of VisiCalc," *BYTE* 5, no. 11 (November 1980): 192; John Markoff, "Radio Shack: Set Apart from the Rest of the Field," *InfoWorld*, July 5, 1982, 43.

42   Gregg Williams and Rob Moore, "The Apple Story," *BYTE*, January 1985, 174.

43   Robert X. Cringely, *Accidental Empires* (Reading, MA: Addison-Wesley, 1991), 71.

44   Mary Brandel, "PC Software Transforms the PC," *Computerworld*, August 2, 1999.

45   Tim Bajarin, "The Application That Birthed The IBM PC," *Forbes*, August 18, 2021, https://www.forbes.com/sites/timbajarin/2021/08/18/the-application-that-birthed-the-ibm-pc/.

46   John F. McMullen and Barbara E. McMullen, "Apple Charts the Course for IBM," *PC Magazine*, February 1984, 126; "IBM 5100 Portable Computer," IBM Archives (IBM), accessed January 27, 2022, https://www.ibm.com/ibm/history/exhibits/pc/pc_2.html.

47   Deborah Wise, "The Colossus Runs, Not Plods: How the IBM PC Came to Be," *InfoWorld*, August 23, 1982; Barry B. Brey, *The Intel Microprocessors*, 8th ed. (Upper Saddle River, NJ: Pearson Prentice Hall, 2009), 5–7.

48   Wise, "How the IBM PC Came to Be."

49   McMullen and McMullen, "Apple Charts the Course for IBM," 127–128.

50   Mark Leon, "Caldera Reopens 'Settled' Suit, Buys DR DOS," *InfoWorld*, July 29, 1996.

51   "Announcement Press Release," IBM Archives (IBM), accessed January 21, 2022, https://www.ibm.com/ibm/history/exhibits/pc25/pc25_press.html.

52   Dan Bricklin, *Bricklin on Technology* (Indianapolis, IN: Wiley, 2009), 361–362.

53   Bricklin, *Bricklin on Technology*, 372–375.

54   Campbell-Kelly and Ceruzzi, "Interview with Dan Bricklin and Bob Frankston," 52; Fylstra, "The Creation and Destruction of VisiCalc," 13.

55   Pollack, "How a Software Winner Went Sour."

56 Pollack, "How a Software Winner Went Sour."

57 Dan Bricklin, "Patenting VisiCalc," accessed January 28, 2022, http://www.bricklin.com/patenting.htm.

58 Pollack, "How a Software Winner Went Sour."

59 Fylstra, "Creation and Destruction of VisiCalc," 14–15; Pollack, "How a Software Winner Went Sour."

60 Pollack, "How a Software Winner Went Sour."

61 James Langdell, "VisiCalc Production Ends," *PC Magazine*, August 6, 1985.

62 Reuell, "A Vision of Computing's Future."

63 "Bricklin Classroom," A Campus Built on Philanthropy (Boston: Harvard Business School), accessed February 1, 2022, https://www.hbs.edu/about/campus-and-culture/campus-built-on-philanthropy/indoor-spaces/Pages/bricklin-classroom.aspx.

64 Melissa Rodriguez Zynda, "The First Killer App: A History of Spreadsheets," *Interactions* 20, no. 5 (September 2013): 68–72, https://doi.org/10.1145/2509224.

65 *U.S. Industrial Outlook 1974* (Washington, DC: U.S. Department of Commerce, 1974), 279; *1980 U.S. Industrial Outlook*, 255.

66 *1990 U.S. Industrial Outlook* (Washington, DC: U.S. Department of Commerce, 1990), 30.7.

67 Martin Hilbert and Priscila López, "The World's Technological Capacity to Store, Communicate, and Compute Information," *Science* 332, no. 6025 (2011): 61, https://doi.org/10.1126/science.1200970.

68 Hilbert and López, "World's Technological Capacity," pt. Supplemental Online Material.

69 Felix Gross, *Calculator Bibliography (Books and Selected Articles)*, 2.8, 2016, 344, https://hhuc.us/2016/files/Speakers/Felix_Gross/Calculator Bibliography HHC 2016.pdf.

70 B. A. Wyld and D. A. Bell, *The Calculator Revolution: How to Make the Most of Your Calculator* (London: White Lion, 1977); James T. Rogers and Bob Korn, *The Calculating Book : Fun and Games with Your Pocket Calculator* (London: Wildwood House, 1976).

EPILOGUE

1   Michael F. McGovern, "Stacks, 'Pacs,' and User Hacks: A Handheld History of Personal Computing," in *Objects and Investigations, to Celebrate the 75th Anniversary of R. S. Whipple's Gift to the University of Cambridge* (Cambridge: Cambridge University Press, 2019), 291, https://doi.org/10.1017/9781108633628.015.

2   Patrick Marshall, Siobhan Nash, and Sebastian Rupley, "Product Comparison," *InfoWorld*, December 1991, 69–81.

3   "IBM Simon," Mobile Phone Museum, accessed April 7, 2022, https://www.mobilephonemuseum.com/phone-detail/ibm-simon; Ben Wood, personal correspondence with the author, 2022.

4   *Nokia 9000 User's Manual* (Nokia Mobile Phones, 1997), 11.1.

5   Gross, *Calculator Bibliography (Books and Selected Articles)*, 344.

6   *Nokia 9000 User's Manual* (Nokia Mobile Phones, 1997), 11:1.

7   Martin Campbell-Kelly, personal correspondence with the author, January 7, 2022.

8   M. D. McIlroy, "A Research UNIX Reader: Annotated Excerpts from the Programmer's Manual, 1971–1986," 1986, 2; D. M. Ritchie, "The UNIX System: The Evolution of the UNIX Time-sharing System," *AT&T Bell Laboratories Technical Journal* 63, no. 8 (1984): 1587, https://doi.org/10.1002/j.1538-7305.1984.tb00054.x.

FURTHER READING

BOOKS

Aspray, William, ed. *Computing Before Computers*. Ames: Iowa State University Press, 1990.

Berlin, Leslie. *The Man Behind the Microchip: Robert Noyce and the Invention of Silicon Valley*. New York: Oxford University Press, 2005.

Bruton, Eric. *The History of Clocks and Watches*. London: Little, Brown, 2000.

Ceruzzi, Paul E. *Reckoners: The Prehistory of the Digital Computer, from Relays to the Stored Program Concept, 1935–1945*. Westport, CT: Greenwood Press, 1983.

Grier, David Alan. *When Computers Were Human*. Princeton: Princeton University Press, 2005.

Ifrah, Georges, and David Bellos. *The Universal History of Numbers: From Prehistory to the Invention of the Computer*. New York: Wiley, 2000.

Kojima, Takashi. *Advanced Abacus: Japanese Theory and Practice*. Rutland, VT: C. E. Tuttle, 1963.

———. *The Japanese Abacus: Its Use and Theory*. Rutland, VT: C. E. Tuttle, 1954.

McCarty, Cara. *Mario Bellini, Designer*. New York: Museum of Modern Art, 1987. https://www.moma.org/calendar/exhibitions/1785.

Menninger, Karl. *Number Words and Number Symbols: A Cultural History of Numbers*. Cambridge: MIT Press, 1969.

Needham, Joseph, and Ling Wang. "Mathematics and the Sciences of the Heavens and the Earth." In *Science and Civilisation in China*, Vol. 3. Cambridge: Cambridge University Press, 1959.

Secrest, Meryle. *The Mysterious Affair at Olivetti: IBM, the CIA, and the Cold War Conspiracy to Shut Down Production of the World's First Desktop Computer.* New York: Alfred A. Knopf, 2019.

Shetterly, Margot Lee. *Hidden Figures: The Untold Story of the African American Women Who Helped Win the Space Race.* London: William Collins, 2017.

PAPERS AND ARTICLES

Aspray, William. "The Intel 4004 Microprocessor: What Constituted Invention?" *IEEE Annals of the History of Computing* 19, no. 3 (1997): 4–15. https://doi.org/10.1109/85.601727.

Cass, Stephen. "Chip Hall of Fame: Intel 4004 Microprocessor." *IEEE Spectrum,* July 2, 2018.

"Chip Hall of Fame: Zilog Z80 Microprocessor." *IEEE Spectrum,* June 2017.

Edelman, Sheldon. "Made in USA . . . Finally!" *The Electronic Engineer,* March 1972. http://www.vcalc.net/hp-35.htm.

Edwards, Paul N., and J.A.N. Lee. "Computer Pioneers." *Technology and Culture* 37, no. 4 (1996). https://doi.org/10.2307/3107118.

Hamrick, Kathy B. "The History of the Hand-Held Electronic Calculator." *American Mathematical Monthly* 102 (1996): 633–639.

Hughes, Jim. "The HP-35." *Codex99,* May 2, 2018. http://codex99.com/design/the-hp35.html.

Mori, Elisabetta. "The Italian Computer: Italy's Olivetti Was an Early Pioneer of Digital Computers and Transistors." *IEEE Spectrum* 56, no. 6 (June 1, 2019): 40–47. https://doi.org/10.1109/MSPEC.2019.8727145.

Okon, Thomas M. "The First Handheld Digital Calculator Celebrates 50 Years, Part 1." *Electronic Design,* March 27, 2017. https://www.electronicdesign.com/technologies/analog/article/21804824/the-first-handheld-digital-calculator-celebrates-50-years-part-1.

———. "The First Handheld Digital Calculator Celebrates 50 Years, Part 2." *Electronic Design,* April 5, 2017. https://www.electronicdesign.com/technologies/analog/article/21804872/the-first-handheld-digital-calculator-celebrates-50-years-part-2.

Reid, T. R. "The Texas Edison." *Texas Monthly,* July 1982. https://www.texasmonthly.com/articles/the-texas-edison/.

Rhines, Walden C. "The Texas Instruments 99/4: World's First 16-Bit Home Computer." *IEEE Spectrum*, June 22, 2017. https://spectrum.ieee.org/the -texas-instruments-994-worlds-first-16bit-computer.

Riordan, Michael. "The Lost History of the Transistor." *IEEE Spectrum* 41, no. 5 (2004): 44–49. https://doi.org/10.1109/MSPEC.2004.1296014.

Stephens, Carlene, and Maggie Dennis. "Engineering Time: Inventing the Electronic Wristwatch." *The British Journal for the History of Science* 33, no. 4 (2000): 477–497.

Stoll, Cliff. "When Slide Rules Ruled." *Scientific American* 294, no. 5 (2006): 80–87.

Thompson, Joe. "The Lost Chapter: A Concise History of the LED Watch." Hodinkee, February 26, 2018. https://www.hodinkee.com/articles/four -revolutions-led-watches.

Wolfe, Tom. "The Tinkerings of Robert Noyce: How the Sun Rose on the Silicon Valley." *Esquire*, 1983. https://www.esquire.com/news-politics/ a12149389/robert-noyce-tom-wolfe/.

INTERVIEWS AND PERSONAL ACCOUNTS

Aspray, William. "Interview with Tadashi Sasaki." IEEE History Center, May 25, 1994. http://ethw.org/Oral-History:Tadashi_Sasaki.

———. "Oral History of Mitch Kapor." Mountain View, CA: Computer History Museum, November 19, 2004. https://www.computerhistory.org/ collections/catalog/102657943.

Baldwin, Frank Stephen. "An Interview with the Father of the Calculating Machine." Monroe Calculating Machine Company, 1919.

Campbell-Kelly, Martin, and Paul Ceruzzi. "An Interview with Dan Bricklin and Bob Frankston (OH 402)." Charles Babbage Institute, 2004. https:// conservancy.umn.edu/handle/11299/113026.

Cochran, David S. "A Quarter Century at HP." HP Memory Project, 2011. https://www.hpmemoryproject.org/timeline/dave_cochran/a_quarter_ century_at_hp_00.htm.

Faggin, Federico, Hal Feeney, Ted Hoff, Stan Mazor, and Masatoshi Shima. "Oral History Panel on the Development and Promotion of the Intel 4004 Microprocessor," 2007. https://www.computerhistory.org/collections/ catalog/102658187.

Fylstra, Dan. "Personal Account: The Creation and Destruction of VisiCalc." May 2004.

Goldstein, Andrew. "Interview with Gordon K. Teal." IEEE History Center, December 1991. https://ethw.org/Oral-History:Gordon_K._Teal.

Merzbach, Uta C. "Interview with Grace Murray Hopper." Computer Oral History Collection, 1969–1973, 1977. Washington, DC: National Museum of American History, 1969.

Minck, John. "Inside HP: A Narrative History of Hewlett-Packard from 1939–1990." HP Memory Project. November 2011. https://www.hpmemoryproject.org/timeline/john_minck/HPNarrative190505.pdf.

Tomash, Erwin. "An Interview with Curt Herzstark (OH 140)." Minneapolis: Charles Babbage Institute, 1987.

Wolff, Michael. "Interview with Jack St. Clair Kilby." IEEE History Center, December 2, 1975. https://ethw.org/Oral-History:Jack_Kilby.

WEBSITES AND COLLECTIONS

Bensene, Rick. "The Old Calculator Web Museum." Accessed March 25, 2022. https://www.oldcalculatormuseum.com/.

Bricklin, Dan. "Dan Bricklin's Web Site." Accessed March 25, 2022. http://www.bricklin.com/.

Devidts, Serge. "Calcuseum." Accessed March 25, 2022. http://www.calcuseum.com/.

Doensen, Pieter. "WATCH: History of the Modern Wrist Watch." Accessed March 25, 2022. https://doensen.home.xs4all.nl/index.html.

Falk, Jim. "Things That Count." Accessed March 25, 2022. http://meta-studies.net/pmwiki/pmwiki.php?n=Site.Introduction.

Hicks, David G. "The Museum of HP Calculators." Accessed March 25, 2022. https://www.hpmuseum.org/.

"IPSJ Computer Museum." Information Processing Society of Japan. Accessed March 25, 2022. https://museum.ipsj.or.jp/en/index.html.

Mislanghe, Marc, ed. "HP Memory Project." Accessed March 25, 2022. https://hpmemoryproject.org/.

"The Oughtred Society." Accessed March 25, 2022. https://www.oughtred.org/.

Tout, Nigel. "Bell Punch Company & Anita Calculators." Accessed March 25, 2022. http://anita-calculators.info/.

————. "Vintage Calculators Web Museum." Accessed March 25, 2022. http://www.vintagecalculators.com/.

Woerner, Joerg. "Datamath Calculator Museum." Accessed March 25, 2022. http://www.datamath.org/.

Wolf, Robert P. "Collection of Slide Rule Replicas." Accessed March 25, 2022. https://www.sliderules.org/.

Wolff, John. "John Wolff's Web Museum—Calculating Machines." Accessed March 25, 2022. http://www.johnwolff.id.au/calculators/.

Zeldes, Nathan. "History of Computing." Accessed March 25, 2022. https://www.nzeldes.com/HOC/HOC_Core.htm.

INDEX